TUTOR WIZARD STATISTICS

Using Microsoft® Excel
Statistics for People Living in the Real World

Nathan H. Johnson, Ph.D.

ABOUT THE AUTHOR

Dr. Nathan Johnson has a Doctorate in toxicology and a Master of Management Degree. He holds advance certifications in clinical chemistry, laboratory management, and hospital administration. Dr. Johnson has been a program manager, principal investigator, and teaches graduate level biostatistics.

The information and advice provided in this book is on an "as is" and "as available" basis. There is no representations or warranties of any kind, expressed or implied. This book is intended to provide information on very basic statistical concepts. The reader is encouraged to consult other sources to include textbooks, software programs, and professional statisticians.

ACKNOWLEDGEMENTS

I would like to thank my Creator for instilling in me a love of numbers. I would like to thank family, friends, and co-workers for listening to my statistical ramblings over the past 25 years. I would also like to thank the inventors of baseball, for it is baseball where my love of numbers blossomed.

Copyright 2014, Nathan Johnson. All Rights Reserved. For bulk ordering, please email - drnathanjohnson@gmail.com
ISBN 978-1-312-77033-1

TABLE OF CONTENTS

ABOUT THE AUTHOR 2

CHAPTER 1: INTRODUCTION 5

CHAPTER 2: BASIC CALCULATIONS USING MICROSOFT® EXCEL 6

CHAPTER 3: GETTING STARTED WITH STATISTICS... 16

CHAPTER 4: PRESENTING DATA USING MICROSOFT® EXCEL 18

CHAPTER 5: CHARACTERIZATION OF DATA 30

CHAPTER 6: BINOMIAL DISTRIBUTIONS 53

CHAPTER 7: POISSON DISTRIBUTIONS 66

CHAPTER 8: INTRODUCTION TO INFERENTIAL STATISTICS 75

CHAPTER 9: NORMAL DISTRIBUTIONS 78

CHAPTER 10: CONFIDENCE INTERVALS 95

CHAPTER 11: t DISTRIBUTION 103

CHAPTER 12: TESTING FOR SIGNIFICANT DIFFERENCES – LARGE SAMPLES..........109

CHAPTER 13: TESTING FOR SIGNIFICANT DIFFERENCES – SMALL SAMPLES..........117

CHAPTER 14: COMPARING TWO MEANS – TWO SAMPLE Z TEST 124

CHAPTER 15: COMPARING TWO MEANS – T TEST FOR RELATED SAMPLES (PAIRED T TEST)... 135

CHAPTER 16: TESTING FOR SIGNIFICANT DIFFERENCES – INDEPENDENT SMALL SAMPLES 144

CHAPTER 17: TESTING FOR SIGNIFICANT DIFFERENCES – MORE THAN TWO GROUPS – ONE WAY ANALYSIS OF VARIANCE... 160

CHAPTER 18: POST-HOC ANALYSIS 174

CHAPTER 19: TESTS FOR DIFFERENCES IN PROPORTIONS 184

CHAPTER 20: CORRELATION AND REGRESSION 194

REFERENCE: STATISTICAL FUNCTIONS USING MICROSOFT® EXCEL 211

TABLE I: The Standard Normal Distribution... 216

TABLE 2: Student's t Distribution 217

CHAPTER 1: INTRODUCTION

In my professional career, both in practice and in the academic world, there are few words that strike fear into the hearts of colleagues and students like that of "statistics". Why is this so? I think I understand many of the reasons. There is the fear of mathematical equations and basic math in general. There is the fear of looking or sounding foolish when discussing statistics. In most cases, the fear is simply of the unknown. Most who fear statistics have never even taken a statistics class. This is a real shame.

The purpose of this book <u>is not</u> to replace a statistics course. A detailed semester long statistics course is beneficial in so many ways. The purpose of this book is to gently introduce the reader to real world statistical problem sets and demonstrate how to use Microsoft® Excel to solve these problems. Each chapter consists of a short introduction of the topic followed by step by step examples with solutions. At the end of each chapter are questions for the reader to test their ability to solve problems. It is hoped this method will help the reader gain confidence in their ability to use statistics as a tool in their everyday life.

CHAPTER 2: BASIC CALCULATIONS USING MICROSOFT® EXCEL

The very first step in using Microsoft® Excel to perform statistical calculations is to understand how to use the program to perform basic mathematical functions. For this chapter and the remainder of this book, Microsoft® Excel 2013 will be used. The reader many have a different version that may require slightly different steps than the ones demonstrated in this textbook.

This introduction assumes the reader has a very basic understanding of Microsoft® Excel. If this is not the case, it is strongly suggested the reader spends a few hours studying a basic Microsoft® Excel self-help book or viewing some of the tutorials on YouTube or other similar resource. After opening or creating a workbook, notice a question mark symbol (?) in the upper right hand corner. This is Microsoft® Excel's help function. This should always be the first stop for any question. Many are surprised by the level of help afforded by this feature. Please refer to Figure 1.

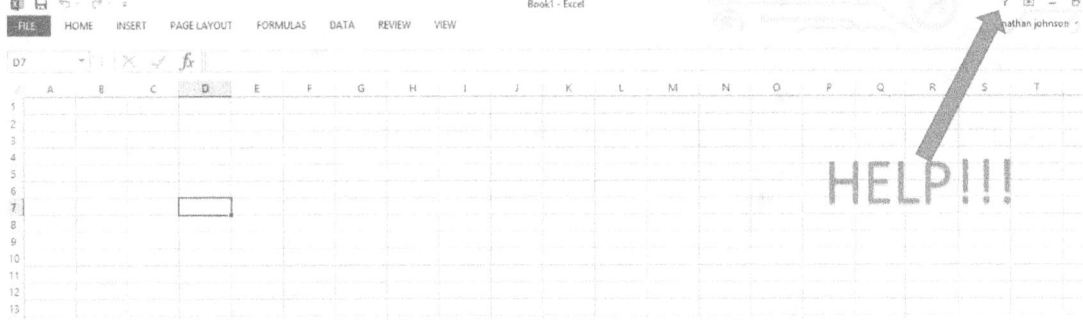

Figure 1: The Help Function

Each entry into an excel workbook is entered into a "cell". The cell is easily identified by its location on the workbook in a manner similar to locating one's position on a worldmap, in which case we use longitude and latitude. In the case of Microsoft® Excel, we use "column letter" and "row number". In Figure 2, $100 is entered in a "cell". This cell is identified by "column letter E" and "row number 5". A shortcut is to use the term "E5" to refer to this cell. Any future reference to a particular cell will be done in this manner. Note that the column and row are both highlighted when the cursor is located in a particular cell. In this example, both the column and row (E5) are highlighted.

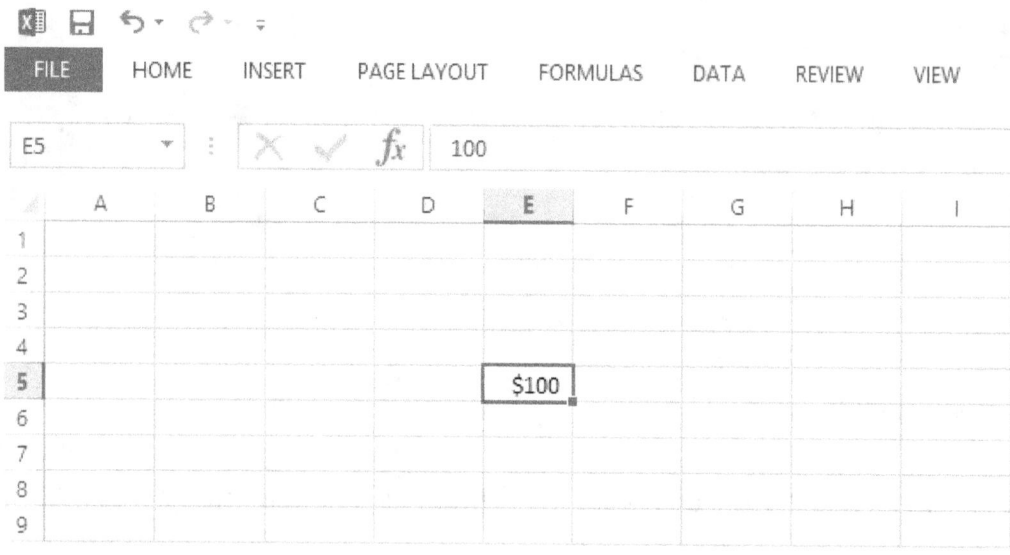

Figure 2: Naming a Cell

In many cases, the desire is to perform calculations on a number of cells. Microsoft® Excel allows this. First, position the cursor in the first cell of interest. Next, using a mouse or trackpad, left click and drag until all cells of interest are highlighted. After releasing the mouse or trackpad, the cells should stay highlighted. This can be done either vertically or horizontally. Please see Figures 3 and 4 for examples.

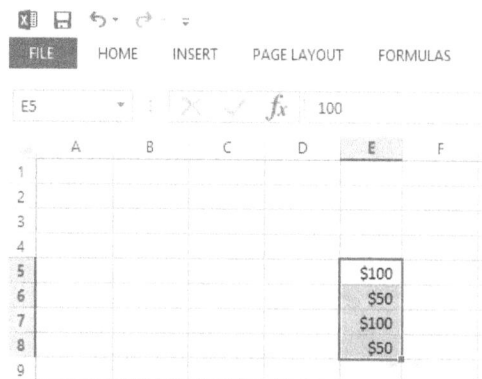

Figure 3: Cells Highlighted Horizontally

Figure 4: Cells Highlighted Vertically

In the remaining Chapters, there will also be many references to the menu bar. The menu bar contains the major items displayed at the top of the workbook. Please refer back to Figure 4 and note the items at the top of the workbook. File, home, formulas, and data will be used on a routine basis.

Although many of the calculations done in Microsoft® Excel are automated, it is still a very good idea to understand how to "program" Microsoft® Excel to perform basic mathematical functions. The basics to using Microsoft® Excel to perform mathematical functions include three simple steps. After placing the cursor in the cell where the result should be displayed, type an equal sign (=). Next, type the abbreviation of the math function that should be performed (e.g. "sum" if the need is to add the items) and finally type the cells that the calculation should be performed on in parentheses. For example, if the desire is to sum the numbers in cells A5 to A10, the following would be typed: =sum (A5:A10) Figure 5 shows some common syntax used.

Operation	Microsoft® Excel Syntax
Add	+
Subtract	-
Divide	/
Multiply	*
Average	Average
Count of Numbers	Count
Maximum Number	Max
Minimum Number	Min
Logarithim	Log
Factorial	Fact
Square Root	SqRt
Standard Deviation	StDev.S

Figure 5: Syntax

Another very helpful thing to understand is how to change the formatting of a cell to produce the output of the calculation. Two common examples are the need to change the number of decimal places in the output or changing a regular number to currency. How would one do that? Let's use this example. Assume a calculation has been performed and the output is as follows (1.354):

Figure 6: Changing Format of Number

Suppose the need is to change the "1.354" into a dollar amount to show "$1.35". Two things need to be done. First, highlight the cell that contains the 1.354 and then left click and a menu will display as seen in Figure 7. Right click on "format cells".

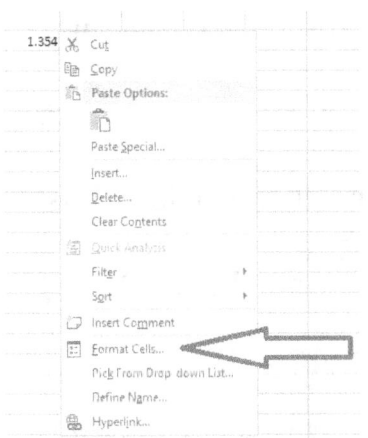

Figure 7: Format Cells

The next thing that should be noticed is a "format cells" option box will appear. See Figure 8:

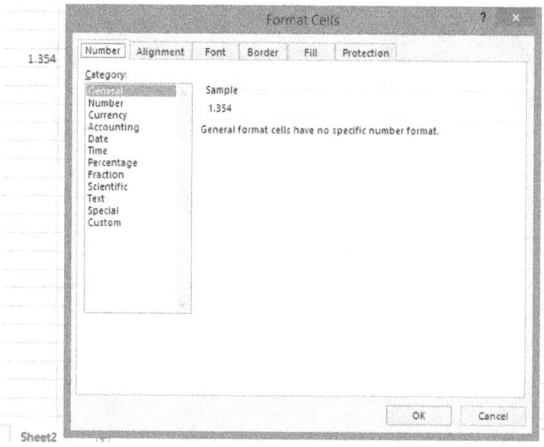

Figure 8: Format Cell Dialogue Box

At this point, the ability to pick one of the categories will be available. In this example, the selection is "number" and change the decimals from three to two. Next choose currency and select OK.

There is one last thing that needs to be done to take full advantage of the statistical power that Microsoft® Excel affords. The Analysis Toolpak needs to be added. The exact way to do this varies with each version of Microsoft® Excel. The example shown is from the 2013 version. First, select "File" from the menu bar and then select "options" as shown in Figure 9:

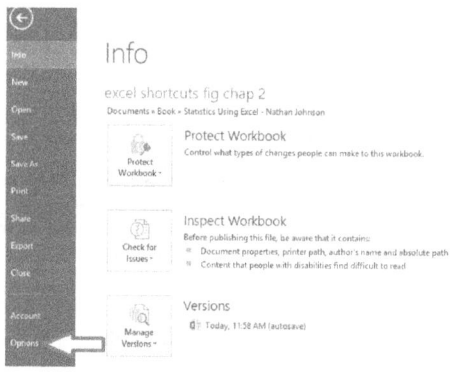

Figure 9: Adding Analysis Toolpak

The next screen seen is a dialogue box that includes all the options that are available. Select Add-Ins

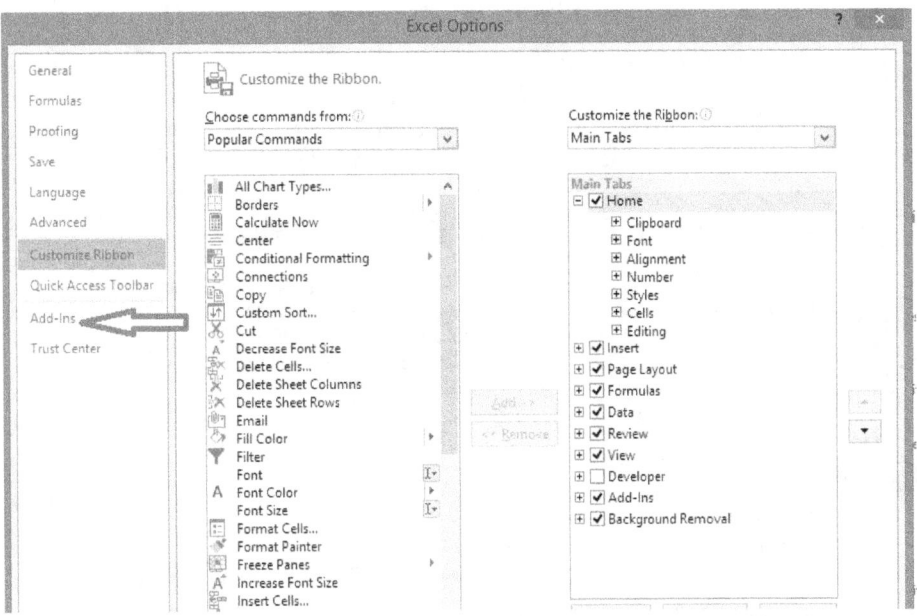

Figure 10: Adding Analysis Toolpak 2

Next a dialogue box will appear that allows the selection of add-ins.

Select Analysis Tool Pack and select "go".

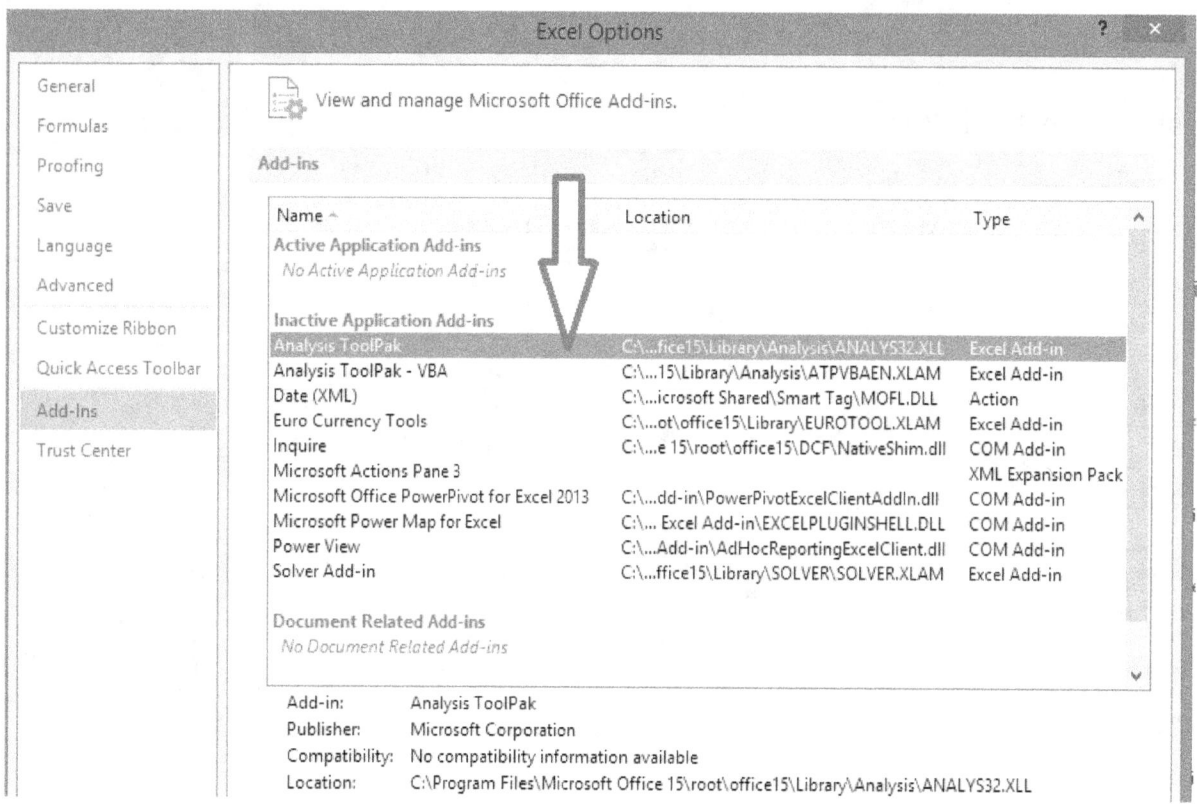

Figure 11: Adding Analysis Toolpak 3

Next, the options for "Add-Ins" will be displayed. Select "Analysis ToolPak" and select "OK". See Figure 12:

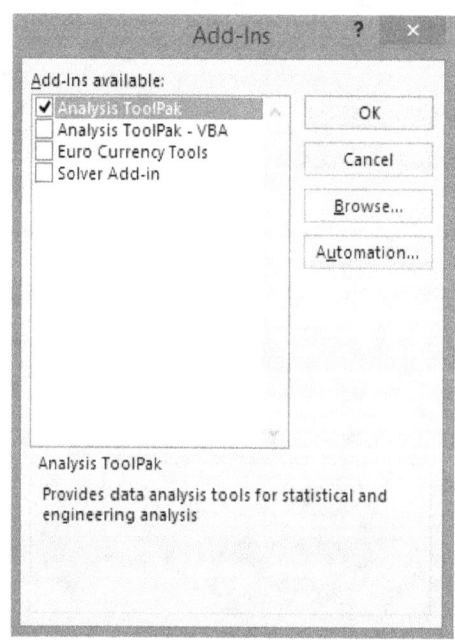

Figure 12: Adding Analysis Toolpak 3

After the Analysis TookPak has been installed, the following icon be seen. It is labeled "Data Analysis" under the data tab. See Figure 13:

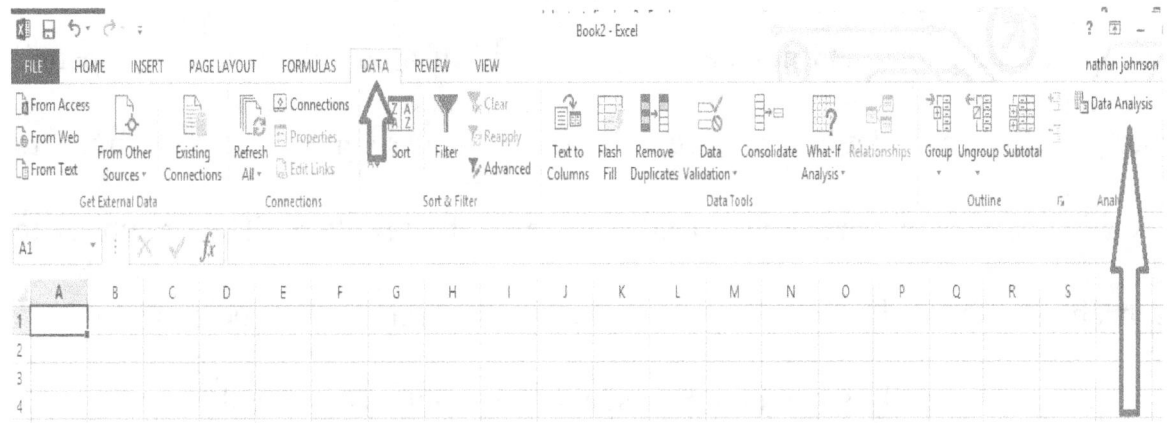

Figure 13: Data Analysis From Menu Bar

CHAPTER 3: GETTING STARTED WITH STATISTICS

Before starting the journey using Microsoft® Excel to answer statistical questions, it must be understood that there are various ways one can get to an answer. Calculations can be performed manually with the aid of a calculator and/or a pencil and paper. Manual calculations may require a statistical reference as an adjunct to look up critical values or other items needed to solve the problem. The drawback to solving problems manually is the real chance that an error will be made in the process. Very few actually solve complex statistical problems manually anymore. It should be noted that learning to perform statistical analysis manually is still very important and gives the student a solid foundation upon which to build on. On the opposite end of the spectrum, there are professional statistical packages that vary in range of price and user friendliness. Those individuals performing statistical analysis on a daily basis will almost certainly use one of these packages. These programs are powerful, can handle both simple and complex datasets, and generally considered to be the gold standard. This text is designed for those individuals who want to confidently perform statistical analysis, but either do not have the resources to purchase a standalone program or who do not perform statistical analysis on a routine basis. It should also be noted that if questions arise, especially when the statistical analysis is important, it is best to consult a professional statistician. Listed on the next page (Figure 14) are some other resources that may be useful in the "statistical journey". This list is not exhaustive, rather, ones that I have experience with or know are of high quality.

Statistical Resource	Link
Number Cruncher Statistical System	http://www.ncss.com/
Power and Sample Size Software	http://www.ncss.com/
Minitab	http://www.minitab.com/en-us/
SAS	http://www.sas.com/en_us/home.html
SPSS	http://www-01.ibm.com/software/analytics/spss/
WINKS	http://www.texasoft.com/
Stat Pages	http://www.statpages.org/
Random.org	http://www.random.org/

Figure 14: Statistical Resources

CHAPTER 4: PRESENTING DATA USING MICROSOFT® EXCEL

One of the first challenges that individuals may experience in their academic pursuits or in the workplace is quickly and accurately depicting data. This has proven to be a challenge to many. Microsoft® Excel makes creating high quality graphical representations of data a breeze.

The first step is to arrange the data in the way the user would like to see it depicted. In some cases, the user may want to show it exactly the way the data is currently formatted. In other cases, the user may want to organize the data in a particular fashion. For example, the user may have the grades of a particular test and want to show the grades from the highest to the lowest. Conversely, the user may want to show the grades from lowest to highest. The way to manipulate the data is the same. The first step is to highlight the data the user wants to organize. The second step, under data in the menu bar, is to select data sort. The user can select highest to lowest, lowest to highest, or sort multiple rows/columns of data keyed to an anchor row/column. The way to sort from highest to lowest and lowest to highest is shown on the next page in Figure 15:

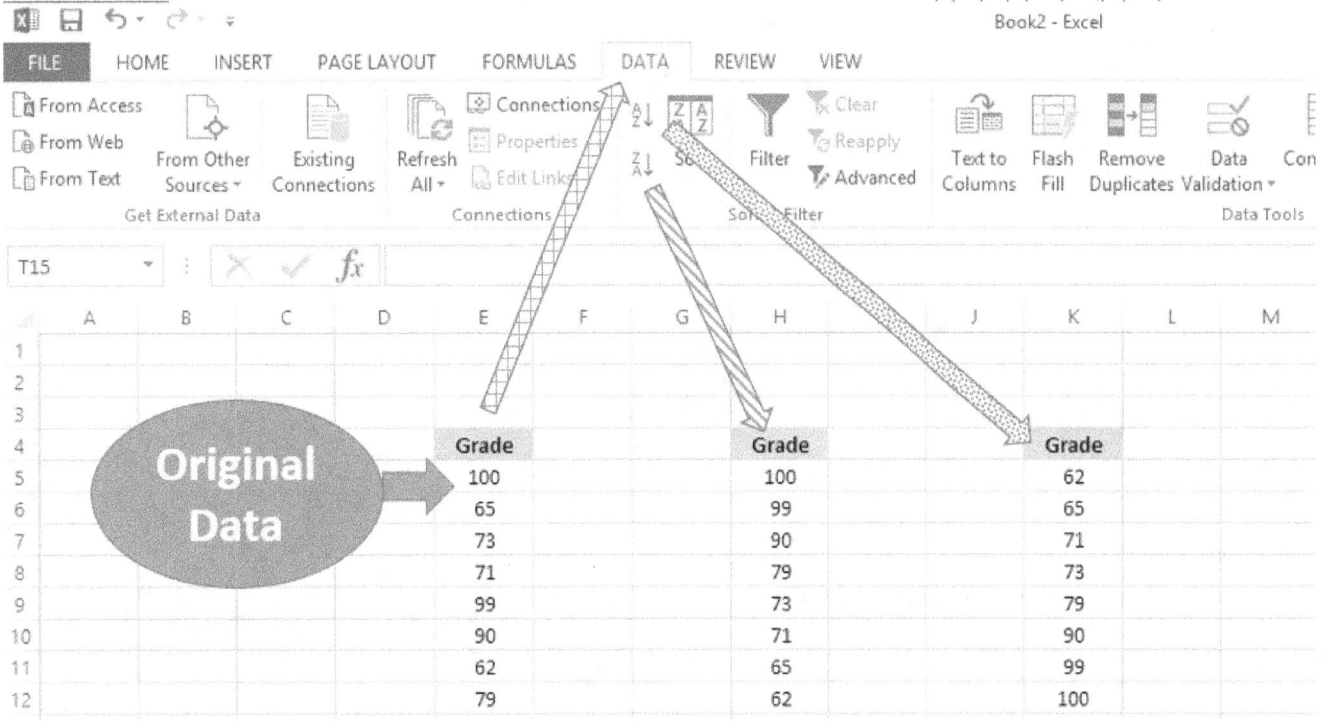

Figure 15: Sorting Data

If the user wants to display any of the data in the cells above, Microsoft® Excel allows that in the click of a few buttons. The more difficult thing is determining exactly what graph or chart should be used. On the next few pages, there are several examples of charts and graphs that may be used. To create a graph, the user should highlight the desired data. On the menu bar, select "insert" and the user will be able to select the desired graph. After the chart or graph is generated, the user can left click on the chart and copy it and paste it into a PowerPoint presentation, a word document, or any other file of the user desires.

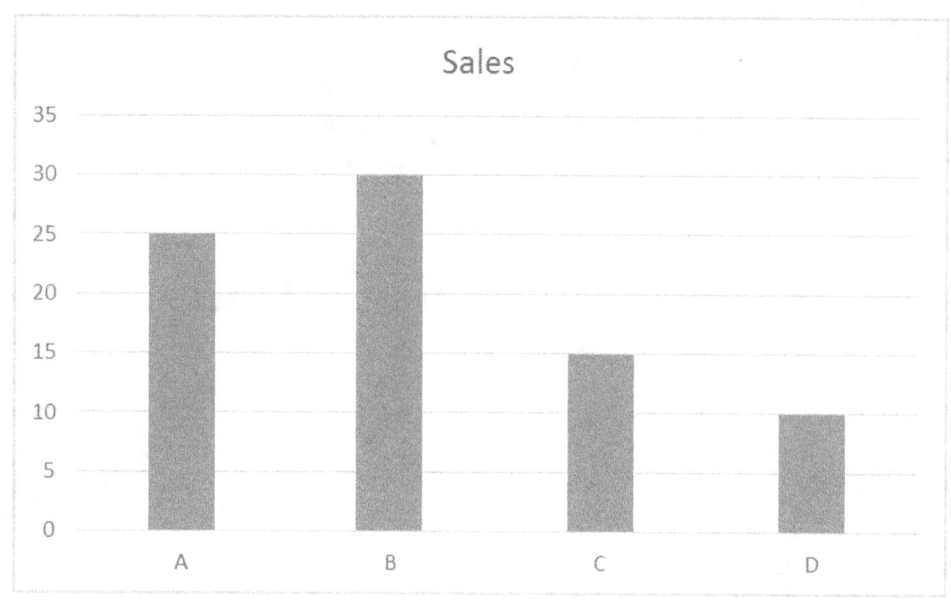

Figure 16: Example Column Chart

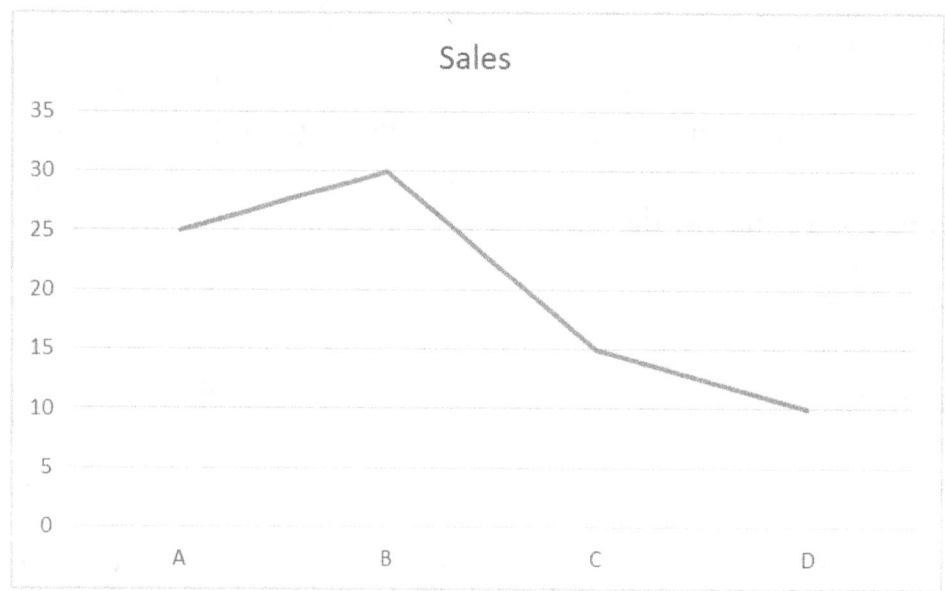

Figure 17: Example Line Chart

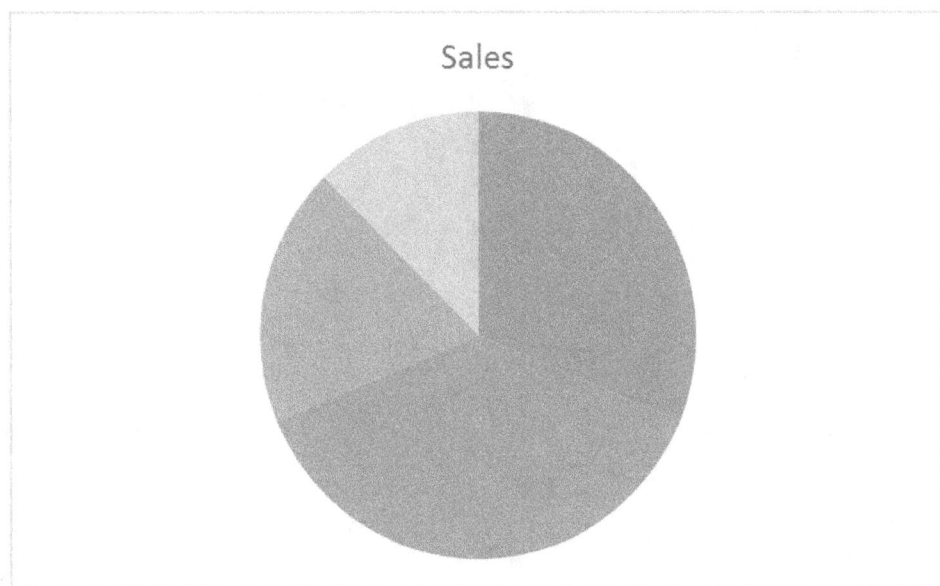

Figure 18: Example Pie Chart

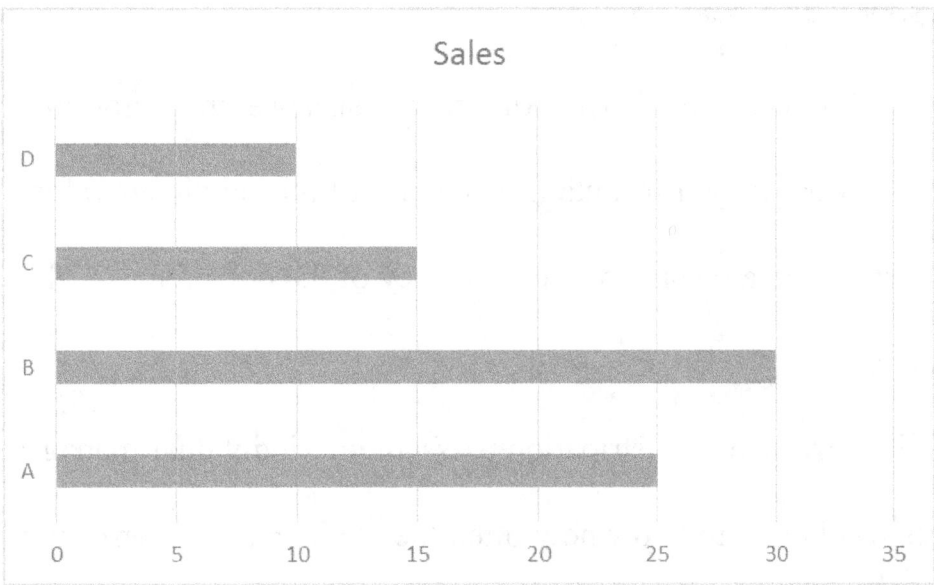

Figure 19: Example Bar Chart

Figure 20: Example Scatter Plot

Microsoft® Excel can also be of great assistance in performing frequency distributions. A frequency distribution is a count of all individual values, or groups of values that results in a display of the frequency of the occurrence of the values of a groups of values.

Example: Frequency Distribution –Assume an individual manages 15 clerks. This manager would like to know how often the clerks make an error per 1,000 files processed. The errors per individual clerk are listed in Figure 21:

Clerk	Errors
1	3
2	5
3	8
4	10
5	20
6	16
7	14
8	2
9	7
10	8
11	9
12	11
13	15
14	13
15	2

Figure 21: Errors Per Clerk

The next step is to decide the groupings or "BINS". Microsoft® Excel recognizes the term BIN to help with this type of problem. How the user sets up the BINS is individual preference, but most find it convenient to divide them into equal sizes. For example, in our example, the lowest number of errors is 2 and the highest is 20. It could easily have 4 groups of 5 or 5 groups of 4. Either would be acceptable. It could even have 2 groups of 10 or it could have unequal size groups. In this example, the problem will solved with 5 groups of 4.

After the data has been entered into a workbook, four more columns will need to be added, the BIN (wording must be exact), errors range, the name of the variable (in our example "# of errors), and relative frequency. The setup should look as follows in Figure 22:

Clerk	Errors	BIN	Errors	# of Errors	Relative Frequency
1	3	4	0 to 4		
2	5	8	5 to 8		
3	8	12	9 to 12		
4	10	16	13 to 16		
5	20	20	17 to 20		
6	16				
7	14				
8	2				
9	7				
10	8				
11	9				
12	11				
13	15				
14	13				
15	2				

Figure 22: Setup of Frequency Distribution – Part I

The steps must be followed in this order. First, highlight the five cells under "# of errors". See Figure 23:

Clerk	Errors	BIN	Errors	# of Errors	Relative Frequency
1	3	4	0 to 4		
2	5	8	5 to 8		
3	8	12	9 to 12		
4	10	16	13 to 16		
5	20	20	17 to 20		
6	16				
7	14				
8	2				
9	7				
10	8				
11	9				
12	11				
13	15				
14	13				
15	2				

Figure 23: Setup of Frequency Distribution – Part II

Next, type "=Frequency (" in the first cell under "# of errors". See Figure 24:

Clerk	Errors	BIN	Errors	# of Errors	Relative Frequency
1	3	4	0 to 4	=frequency(	
2	5	8	5 to 8	FREQUENCY(data_array, bins_array)	
3	8	12	9 to 12		
4	10	16	13 to 16		
5	20	20	17 to 20		
6	16				
7	14				
8	2				
9	7				
10	8				
11	9				
12	11				
13	15				
14	13				
15	2				

Figure 24: Setup of Frequency Distribution – Part III

All the cells under the first "errors" should be highlighted. The user will now see those cells highlighted listed. This should look like the shown in Figure 25:

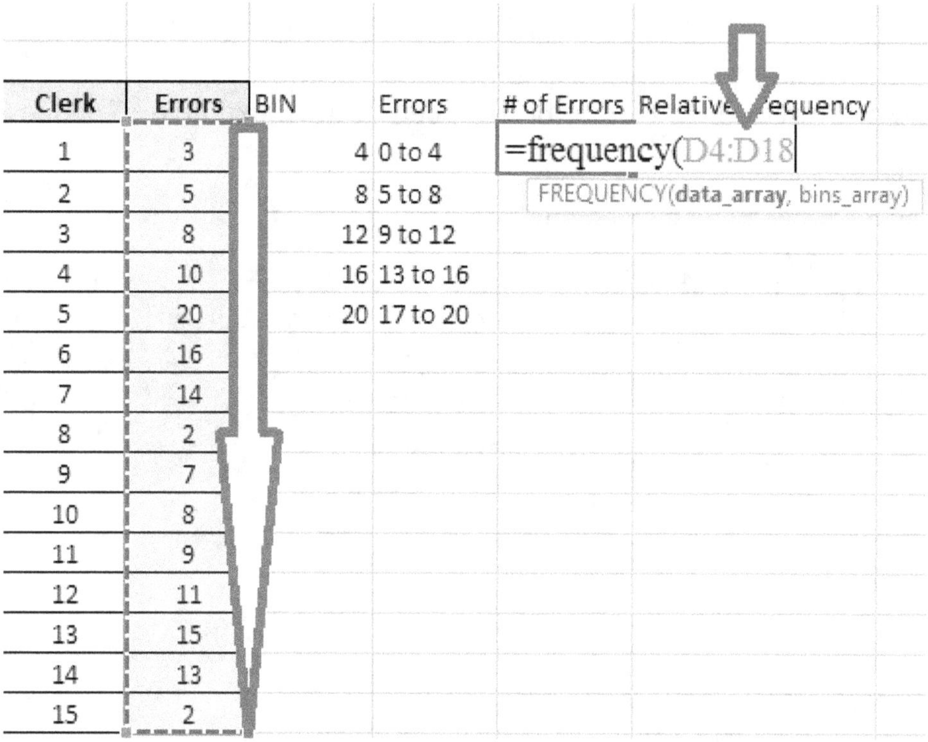

Clerk	Errors	BIN	Errors	# of Errors	Relative Frequency
1	3	4	0 to 4	=frequency(D4:D18	
2	5	8	5 to 8	FREQUENCY(data_array, bins_array)	
3	8	12	9 to 12		
4	10	16	13 to 16		
5	20	20	17 to 20		
6	16				
7	14				
8	2				
9	7				
10	8				
11	9				
12	11				
13	15				
14	13				
15	2				

Figure 25: Setup of Frequency Distribution – Part IV

Next, type a comma after the highlighted cells (in this example, after D:18), and then highlight the cells under "BIN". See Figure 26:

Clerk	Errors	BIN	Errors	# of Errors	Relative Frequency
1	3	4	0 to 4	=frequency(D4:D18,E4:E8	
2	5	8	5 to 8	FREQUENCY(data_array, bins_array)	
3	8	12	9 to 12		
4	10	16	13 to 16		
5	20	20	17 to 20		
6	16				
7	14				
8	2				
9	7				
10	8				
11	9				
12	11				
13	15				
14	13				
15	2				

Figure 26: Setup of Frequency Distribution – Part V

Next, close the parentheses and press simultaneously press control, shift, and enter at the same time. The user will see the cells in the "# of errors" column fill in. See Figure 27.

Clerk	Errors	BIN	Errors	# of Errors	Relative Frequency
1	3	4	0 to 4	3	
2	5	8	5 to 8	4	
3	8	12	9 to 12	3	
4	10	16	13 to 16	4	
5	20	20	17 to 20	1	
6	16				
7	14				
8	2				
9	7				
10	8				
11	9				
12	11				
13	15				
14	13				
15	2				

Figure 27: Setup of Frequency Distribution – Part VI

Next, highlight the cells under the # of errors to include the empty cell under the last number and click autosum. The user will see the numbers are summed. "15": should be the number seen, which is the total number of clerks. See Figure 28.

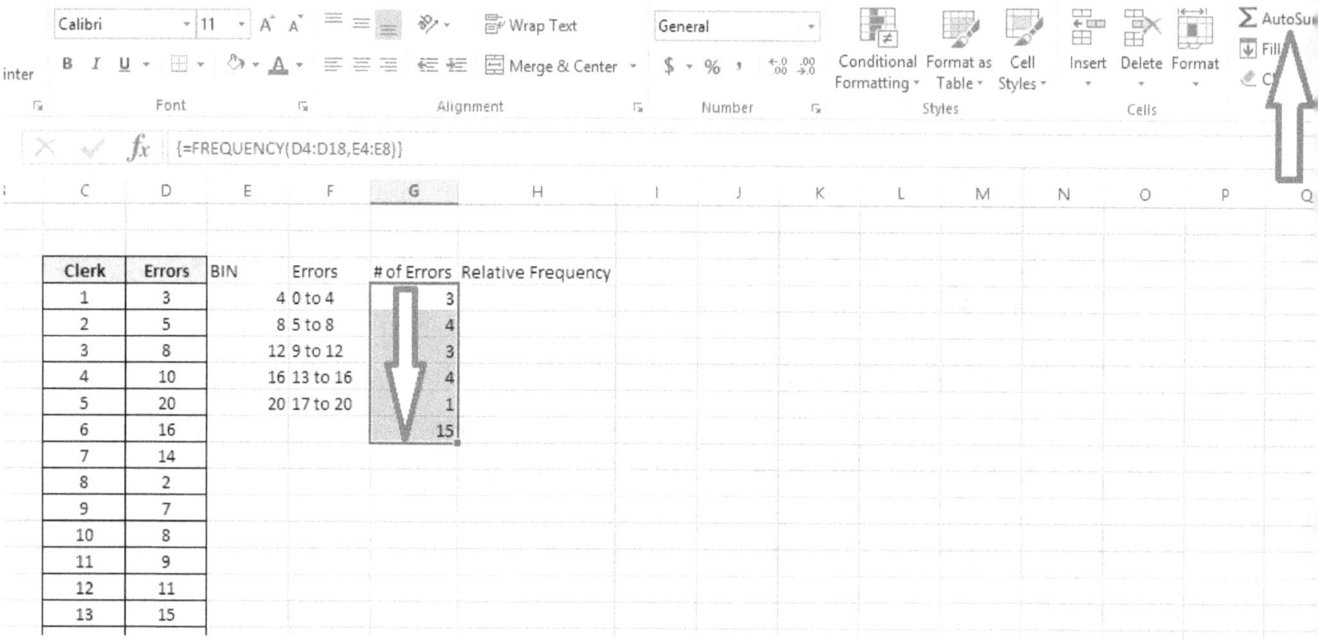

Figure 28: Setup of Frequency Distribution – Part VII

Next, the relative frequency can be calculated. What is relative frequency? It is the percentage of each BIN relative to the total number of observations. In the first cell under "relative frequency", there are two ways this can be solved. One way is just to type "=3/15" and then repeat for each cell. Alternatively, one could type "=cell/$cellcolum$cell row". In this case, it would be appropriate to type "=g4/g9" The user can then cut and paste into each of the remaining cells. The autosum can be used for the number of errors which should add up to one. See Figure 29:

Clerk	Errors	BIN	Errors	# of Errors	Relative Frequency
1	3	4	0 to 4	3	0.2
2	5	8	5 to 8	4	0.266666667
3	8	12	9 to 12	3	0.2
4	10	16	13 to 16	4	0.266666667
5	20	20	17 to 20	1	0.066666667
6	16			15	1
7	14				
8	2				
9	7				
10	8				
11	9				
12	11				
13	15				
14	13				
15	2				

Figure 29: Setup of Frequency Distribution – Part VIII

CHAPTER 5: CHARACTERIZATION OF DATA

Characterization of data is very important to the user of statistics. Generally, in peer reviewed journals or in professional presentations, the first data presented are used to describe the overall nature of the data. In other words, this summary is descriptive in nature. We will refer to this summary as descriptive statistics. The user may find or see different inclusions for descriptive statistics, however, we will cover the most frequently used products.

Before we get into the details, we will take a look at the "end product" of excel and then address each of the items in some detail. Microsoft® Excel can produce descriptive statistics from simple or complex datasets. For our example, we will use the very simple dataset found below which includes test grades on a statistics exam. See Figure 30:

STATISTICS GRADES
95
58
75
77
91
57
80
78
76
88

Figure 30: Test Grades on Statistics Exam

A casual look at the data can lead to some observations. There are not that many observations and they could be counted if desired. It looks like two students made grades above 90 and two made grades below 60. It looks like there is some variation in the grades. This quick summation is actually a form of descriptive statistics. This quick summation is also lacking in many ways. How much variation exists? What is the mean? What does the distribution look like? Microsoft® Excel can quickly answer these and other questions. From the menu bar, select Data Analysis under Data then select "OK". See Figure 31:

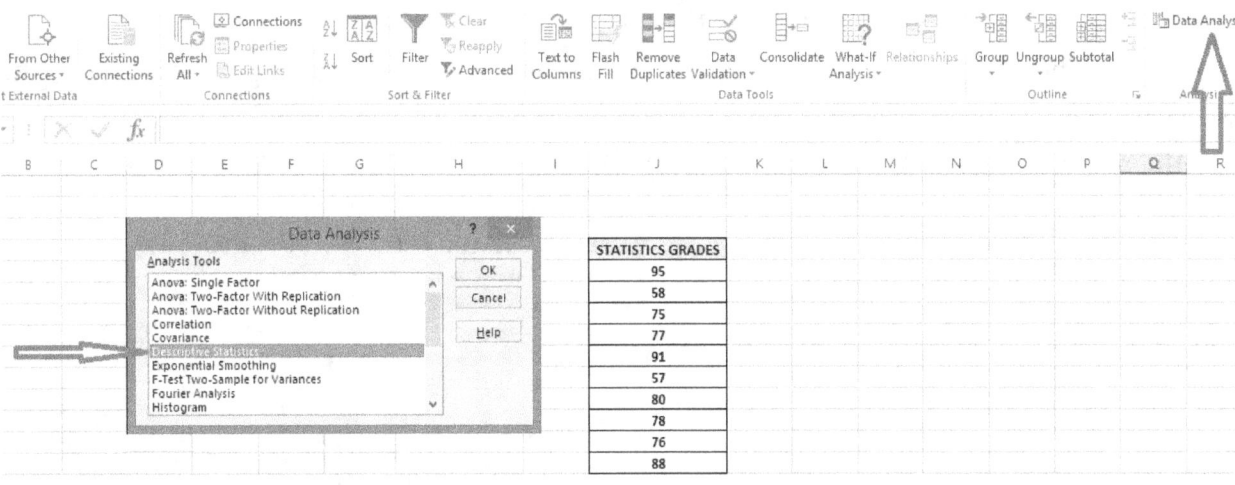

Figure 31: Descriptive Stats I

Next, the following dialogue box can been seen: See Figure 32:

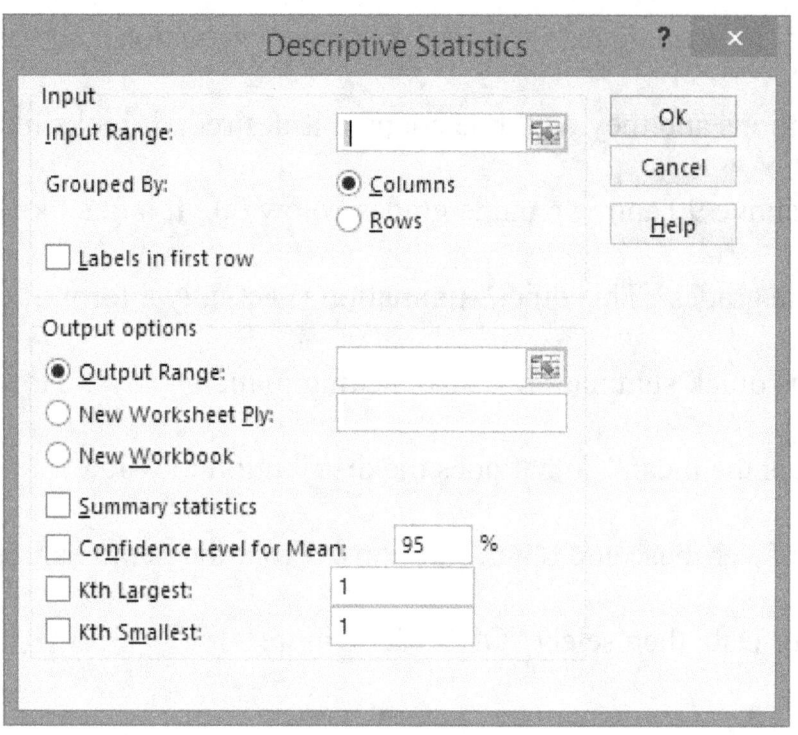

Figure 32: Descriptive Stats II

The user will need to position the cursor in the "input range" box and highlight the cells of interest. In this case, all the grades are highlighted. The user will now see the same dialogue box with the cell range displayed. See Figure 33:

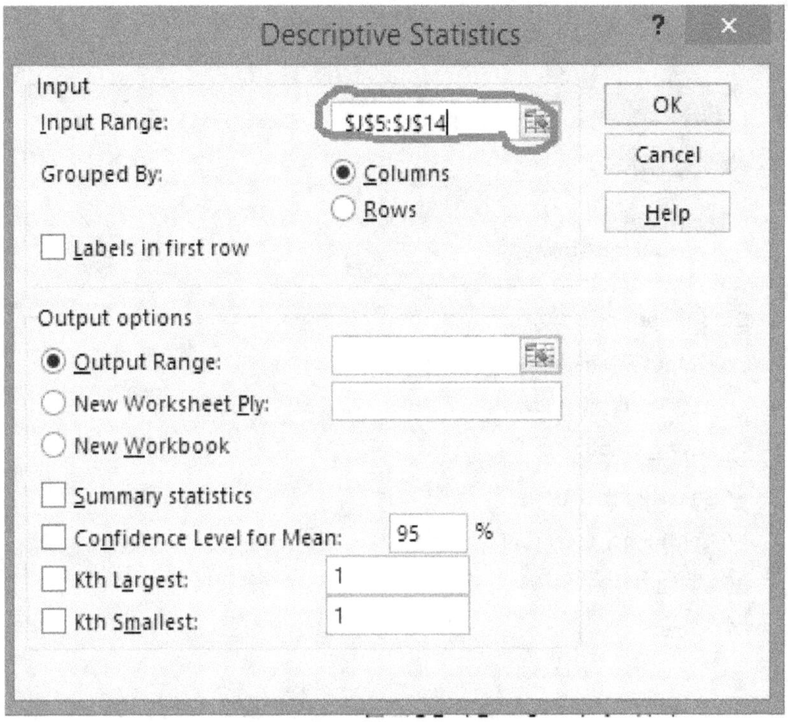

Figure 33: Descriptive Stats III

Next, position the cursor in "output range" and select the desired cell in which the descriptive statistics summary should display. Next, select an area that does not mask the data. For example, the user may choose a cell three or four columns to the left or right of the data. One could also choose to display it below the data. The user should also check the boxes next to "summary statistics", "confidence level for the mean", "Kth largest", and "Kth smallest". The dialogue box should look as shown in Figure 34:

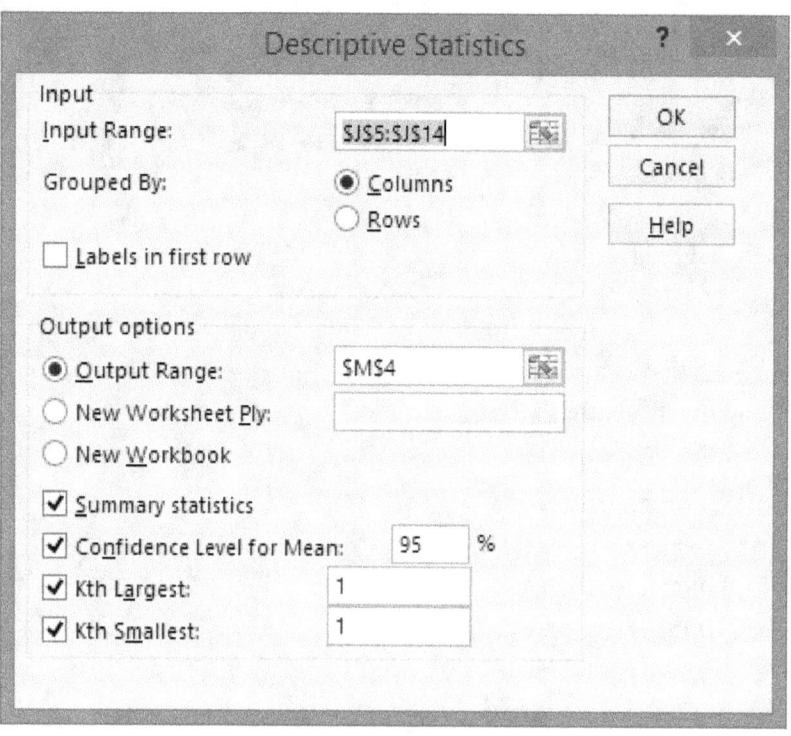

Figure 34: Descriptive Stats IV

Next, the user should press OK. The summary statistics will display in the identifed cell. Please note that it may be necessary to expand the column widith to display all of the text. See Figure 35:

	Column1
Mean	77.5
Standard Error	3.964425137
Median	77.5
Mode	#N/A
Standard Deviation	12.53661305
Sample Variance	157.1666667
Kurtosis	-0.27094998
Skewness	-0.49737667
Range	38
Minimum	57
Maximum	95
Sum	775
Count	10
Largest(1)	95
Smallest(1)	57
Confidence Level(95.0%)	8.96815272

Figure 35: Descriptive Stats V

What do all of these summary statistics mean? A user-friendly definition of each is shown following this paragraph. Each definition will be expanded on in later chapters.

Definitions:

<u>Mean</u>: The sum of the items measured divided by the number of items.

<u>Standard Error</u>: The standard deviation of the population divided by the square root of the sample size. It is a measure of random variability.

<u>Median</u>: The "middle" of the dataset. Half of the items are above and half of the items are below. If there are an even number of observations, the median is the average of the middle two observations.

Mode: The most frequently occurring item in the dataset. If no one observation outnumbers the other numbers in the dataset, the mode is N/A.

Standard Deviation: The square root of variance; this is the overall measure of data variation.

Sample Variance: The mean of squared deviation scores from the means of a distribution. It is calculated by adding the sums of the squared differences of observations from the mean and dividing by the total number of observations minus one.

Kurtosis: Measures the steepness or "peakedness" of the distribution.

Skewness: Measures the symmetry around the mean.

Range: The difference between the lowest and highest measurements.

Minimum: The lowest measurement.

Maximum: The highest measurement.

Sum: The total sum of all observations.

Count: The total number of items measured.

There are three calculations that are crucial to understanding their relationship with each other. Although Microsoft® Excel performs these calculations from the data analysis function, it is important to understand that these can also be manually performed in Microsoft® Excel and learning how to do that will greatly foster a better understanding of their overall relationship. The three calculations are variance, standard deviation, and standard error.

The calculation of variance is fairly simple. We will use the same example of student grades. The first step is to set up the data. In this case, it is the student grades. See Figure 36:

STATISTICS GRADES
95
58
75
77
91
57
80
78
76
88

Figure 36: Student Grades

Next, determine the mean. In this case, the descriptive statistics function was used to determine the mean was 77.5. Another way to obtain the same result is to add the grades and divided by the number of counts, which in this case was 10.

The next step is to determine the deviation for each data point. In other words, how much it differs from the mean. The user can use Microsoft® Excel to perform this simple math. See Figure 37:

STATISTICS GRADES	Difference From Mean
95	17.5
58	-19.5
75	-2.5
77	-0.5
91	13.5
57	-20.5
80	2.5
78	0.5
76	-1.5
88	10.5
Mean 77.5	

Figure 37: Difference From the Mean

Next, square each deviation. In other words, the user needs to multiply the difference by itself. Notice that any negative numbers are now positive. A negative number multiplied by a negative number results in a positive number. The user can use Microsoft® Excel to square the differences. In the first cell, type the cell reference followed by the "^" sign followed by 2. In this case, the formula was "=G4^2". If the user copies this formula and pastes it into the remaining cells, the calculation will automatically be performed. Alternatively, the user could type "=17.5^2" or ="17.5*17.5", but in that case, it would be needed to do that for each individual cell. In large datasets, this may not be a viable alternative. See Figure 38.

STATISTICS GRADES	Difference From Mean	Difference Squared
95	17.5	306.25
58	-19.5	380.25
75	-2.5	6.25
77	-0.5	0.25
91	13.5	182.25
57	-20.5	420.25
80	2.5	6.25
78	0.5	0.25
76	-1.5	2.25
88	10.5	110.25
Mean 77.5		

Figure 38: Differences Squared

Next, calculate the sum of squares (SS). See Figure 39:

	STATISTICS GRADES	Difference From Mean	Difference Squared
	95	17.5	306.25
	58	-19.5	380.25
	75	-2.5	6.25
	77	-0.5	0.25
	91	13.5	182.25
	57	-20.5	420.25
	80	2.5	6.25
	78	0.5	0.25
	76	-1.5	2.25
	88	10.5	110.25
Mean	77.5		1414.5

Figure 39: Sum of Differences Squared

Next, divide the sum of squares (SS) by n-1. It has not been mentioned before, but n is shorthand for the total number of items counted. In this case, n = 10, so n-1 = 9. In this example, we divide 1414.5 by 9 and the result is 157.1667. This is the calculated variance. Please refer back to the descriptive statistics that Microsoft® Excel calculated. The variance is exactly the same.

Next, take the square root of the variance. This is the standard deviation. In Microsoft® Excel, the square root can be calculated by taking the number and taking it to the .5 power. In Microsoft® Excel, this would be taking the square root of the variance, in our case 157.17 and it would be input as "=157.17^.5". See Figure 40:

	STATISTICS GRADES	Difference From Mean	Difference Squared
	95	17.5	306.25
	58	-19.5	380.25
	75	-2.5	6.25
	77	-0.5	0.25
	91	13.5	182.25
	57	-20.5	420.25
	80	2.5	6.25
	78	0.5	0.25
	76	-1.5	2.25
	88	10.5	110.25
Mean	77.5		1414.5
		Variance	157.1666667
		Standard Deviation	12.53661305

Figure 40: Square Root of Variance (Standard Deviation)

Again, the calculation for standard deviation is the same as what Microsoft® Excel calculated for us. To finalize this discussion, calculate the Standard Error of the Mean (SEM). The SEM is calculated by taking the standard deviation and dividing by the square root of the sample size. The SEM in our example is calculated by dividing the standard deviation (12.54) by the square root of the sample size (square root of 10). In this case, the input was "=D16/SQRT (10)". Again, our result is exactly the same as what Microsoft® Excel calculated for us. See Figure 41:

	STATISTICS GRADES	Difference From Mean	Difference Squared
	95	17.5	306.25
	58	-19.5	380.25
	75	-2.5	6.25
	77	-0.5	0.25
	91	13.5	182.25
	57	-20.5	420.25
	80	2.5	6.25
	78	0.5	0.25
	76	-1.5	2.25
	88	10.5	110.25
Mean	77.5		1414.5
		Variance	157.1666667
		Standard Deviation	12.53661305
		SEM	3.964425137

Figure 41: Standard Error of the Mean

Another quick and useful statistical calculation that allows a quick comparison of the variation between datasets is the coefficient of variation. Comparison of standard deviation alone does not account for the possible difference in the means. For example, a standard deviation of 3 for with a mean of 10 is much different than a standard deviation of 3 with a mean of 100. The later has much less variation. The coefficient of variation compensates for the differences in means and is a much better comparison tool. It is calculated by dividing the standard deviation by the mean and multiplying that product by 100. The product of this multiplication is a percentage. For example, if the standard deviation was 10 and the mean was 100, the calculation would be (10/100)*100 or 10.0%. The higher the number, the larger the variation.

Chapter 5 Exercises:

1. Determine the variance, standard deviation, and standard error of the mean on the following dataset. Compare it to what Microsoft® Excel generated using the descriptive statistics function.

362
1
187
584
418
496
139
249
960
327

2. Determine the variance, standard deviation, and standard error of the mean on the following dataset. Compare it to what Microsoft® Excel generated using the descriptive statistics function.

259
796
873
168
439
963
859
648
172
417

3. Determine the variance, standard deviation, and standard error of the mean on the following dataset. Compare it to what Microsoft® Excel generated using the descriptive statistics function.

879
545
662
508
233
426
676
211
411
214

4. Determine the variance, standard deviation, and standard error of the mean on the following dataset. Compare it to what Microsoft® Excel generated using the descriptive statistics function.

543
214
332
61
712
978
879
182
350
964

5. Determine the variance, standard deviation, and standard error of the mean on the following dataset. Compare it to what Microsoft® Excel generated using the descriptive statistics function.

216
340
880
996
774
435
450
834
115
792

6. Determine the variance, standard deviation, and standard error of the mean on the following dataset. Compare it to what Microsoft® Excel generated using the descriptive statistics function.

575	549
734	6
196	572
361	928
253	67
854	918
864	151
511	380

7. Determine the variance, standard deviation, and standard error of the mean on the following dataset. Compare it to what Microsoft® Excel generated using the descriptive statistics function.

426	603
766	586
582	329
655	738
804	486
582	388
229	772
368	410

8. Determine the variance, standard deviation, and standard error of the mean on the following dataset. Compare it to what Microsoft® Excel generated using the descriptive statistics function.

669	910
669	483
536	923
850	278
631	226
123	901
860	93
30	389

9. Determine the variance, standard deviation, and standard error of the mean on the following dataset. Compare it to what Microsoft® Excel generated using the descriptive statistics function.

70	739
316	85
332	470
859	974
121	701
500	401
740	174
951	793

10. Determine the variance, standard deviation, and standard error of the mean on the following dataset. Compare it to what Microsoft® Excel generated using the descriptive statistics function.

547	20
846	124
497	822
64	602
591	904
282	898
563	921
987	132

11. For the first three datasets, calculate the coefficient of variation of the dataset.

Chapter 5 Exercises: Solutions

1.

Column1	
Mean	372.3
Standard Error	85.08741257
Median	344.5
Mode	#N/A
Standard Deviation	269.0700239
Sample Variance	72398.67778
Kurtosis	1.69527197
Skewness	1.014160342
Range	959
Minimum	1
Maximum	960
Sum	3723
Count	10
Largest(1)	960
Smallest(1)	1
Confidence Level	192.4810998

2.

Column1	
Mean	559.4
Standard Error	96.78810763
Median	543.5
Mode	#N/A
Standard Deviation	306.0708705
Sample Variance	93679.37778
Kurtosis	-1.797215596
Skewness	-0.060184737
Range	795
Minimum	168
Maximum	963
Sum	5594
Count	10
Largest(1)	963
Smallest(1)	168
Confidence Level	218.9499109

3.

Column1	
Mean	476.5
Standard Error	70.50851012
Median	467
Mode	#N/A
Standard Deviation	222.9674864
Sample Variance	49714.5
Kurtosis	-0.604370828
Skewness	0.353751924
Range	668
Minimum	211
Maximum	879
Sum	4765
Count	10
Largest(1)	879
Smallest(1)	211
Confidence Level	159.5013312

4.

Column1	
Mean	521.5
Standard Error	108.318281
Median	446.5
Mode	#N/A
Standard Deviation	342.5324802
Sample Variance	117328.5
Kurtosis	-1.663521241
Skewness	0.195878999
Range	917
Minimum	61
Maximum	978
Sum	5215
Count	10
Largest(1)	978
Smallest(1)	61
Confidence Level	245.0329752

5.

Column1	
Mean	583.2
Standard Error	97.42778522
Median	612
Mode	#N/A
Standard Deviation	308.0937087
Sample Variance	94921.73333
Kurtosis	-1.565501051
Skewness	-0.208187553
Range	881
Minimum	115
Maximum	996
Sum	5832
Count	10
Largest(1)	996
Smallest(1)	115
Confidence Level	220.3969622

6.

Mean	494.9375
Standard Error	77.02458854
Median	530
Mode	#N/A
Standard Deviation	308.0983542
Sample Variance	94924.59583
Kurtosis	-1.250529518
Skewness	-0.032409333
Range	922
Minimum	6
Maximum	928
Sum	7919
Count	16
Largest(1)	928
Smallest(1)	6
Confidence Level(95.0%)	164.1740242

7.

Mean	545.25
Standard Error	43.91938638
Median	582
Mode	582
Standard Deviation	175.6775455
Sample Variance	30862.6
Kurtosis	-1.053876429
Skewness	-0.098298496
Range	575
Minimum	229
Maximum	804
Sum	8724
Count	16
Largest(1)	804
Smallest(1)	229
Confidence Level(95.0%)	93.61195614

8.

Mean	535.6875
Standard Error	78.54880217
Median	583.5
Mode	669
Standard Deviation	314.1952087
Sample Variance	98718.62917
Kurtosis	-1.374085348
Skewness	-0.261372133
Range	893
Minimum	30
Maximum	923
Sum	8571
Count	16
Largest(1)	923
Smallest(1)	30
Confidence Level(95.0%)	167.4228087

9.

Mean	514.125
Standard Error	78.38366512
Median	485
Mode	#N/A
Standard Deviation	313.5346605
Sample Variance	98303.98333
Kurtosis	-1.434156273
Skewness	-0.023027566
Range	904
Minimum	70
Maximum	974
Sum	8226
Count	16
Largest(1)	974
Smallest(1)	70
Confidence Level(95.0%)	167.0708274

10.

Mean	550
Standard Error	83.89283839
Median	577
Mode	#N/A
Standard Deviation	335.5713536
Sample Variance	112608.1333
Kurtosis	-1.333826382
Skewness	-0.334931401
Range	967
Minimum	20
Maximum	987
Sum	8800
Count	16
Largest(1)	987
Smallest(1)	20
Confidence Level(95.0%)	178.8133523

11. For problem 1, the CV is 72.3%. For problem 2, the CV is 54.7%. For problem 3, the CV is 46.8%.

CHAPTER 6: BINOMIAL DISTRIBUTIONS

Starting with this Chapter, specific distributions of data will be discussed. In the last Chapter, several key descriptive statistical observations were examined. The next step in this statistical journey will involve learning how to use the data and make critical decisions or judgments on the data.

What are distributions of data? Simply put, it is how the data is situated around any reference point (e.g. the mean). The first question to ask is if the data is continuous or discrete? Continuous data can take on any value. For example, the percentage of United States Citizens who have the last name Johnson. The answer could be 3.1394 percent or it could be 2.9784 percent and so on. If one asked the question "how many of 100 fish caught were bass?" The answer could be any number between 0 and 100, but it could not be 10.5, because one can't catch half a bass. This type of data is discrete. This Chapter will discuss two major discrete distributions that Microsoft® Excel can help with. Continuous data will be discussed in subsequent chapters.

The binomial distribution is the first discrete distribution that will be covered. It is very useful and applicable in everyday life. For data to be considered appropriate for the binomial distribution, several characteristics or properties must me observed. First, the data must have binary outcomes. Examples of a binary outcome are success/failure, yes/no, right/wrong, etc... Second, the outcome of separate observations must be independent. In other words, the outcome of one observation may not impact another observation. For example, the result of the first flip of a coin does impact the result of the second flip of a coin. Lastly, the probability of an observation remains constant. For example, the probability of flip of a coin being "heads" is always .50. When performing calculations using the binomial distribution, the total number of observations must be counted and the probability (*p*) of one outcome must be known. By definition, if *p* is known, then the probability of the alternate observation is 1-*p*. For example, if the probability of a "heads" on a coin flip is .50, then the probability of a "tails" is 1-*p* or 1-.50 which is .50.

The reader should refer to statistical reference texts for a complete discussion of the properties derivations of binomial equations. However, to somewhat understand how Microsoft® Excel is computing solutions to binominal problems, it is a good idea to recognize the basic binomial formula. The probability of an unknown number of observations given a series of trials is defined by the following formula. Although this formula may look daunting, it is actually quite simple. See Figures 42 and 43.

$$P(X) = \frac{n!}{X!(n-X)!} p^X (1-p)^{n-X}$$

Figure 42: Binomial Distribution

	Binomial Distribution - Parts of the Equation
P(X)	probability of X number of observations
n	total number of observations or sample size
p	probability of observation
1-p	probability of alternate observation
X	number of observations
!	factorial: (0=1, 1=1x1, 2=2x1, 3=3x2x1, 4=4x3x2x1, etc...)

Figure 43: Components of the Binomial Equation

A good example of the binomial distribution is a coin flip. Imagine that one flips a coin ten times. What is the probability of getting 5 tails? Using the binomial formula, one would get:

$$P(10) = \frac{10!}{5!(10-5)!} .5^{5}(1-.5)^{10-5}$$

Although we are not using Microsoft® Excel for this example, it may save time to use a programmable calculator or even Microsoft® Excel to solve this problem. In Microsoft® Excel, input "=FACT (10)" to get the factorial for 10. Likewise, to take something to a certain power, as discussed in a previous chapter, input "=.5^5" to take .5 to the 5th power. The summation of this example is 252*0.03125*0.03125 which is 0.246. In other words, there is almost a one in four chance of getting exactly five tails. What if one wanted to determine what the probability of getting two heads? Using the binomial formula, the following can be calculated:

$$P(2) = \frac{10!}{2!(10-2)!} .5^2 (1-.5)^{10-2}$$

The summation of this example is 45*0.25*0.0039 which is 0.044. In other words, there is less than a five percent probability of getting exactly two heads.

Microsoft® Excel can easily do the same calculations all at once. The first step is to set up the workbook in Microsoft® Excel. Using coin flip example, we can easily and quickly determine not only the probability of 5 heads but also the probability of every other outcome. For this example, the problem is setup like this:

	probability	cumulative probability
0		
1		
2		
3		
4		
5		
6		
7		
8		
9		
10		

Figure 44: Binomial Setup Part I

Next, position the cursor in the first cell. In this example, it is the cell adjacent to the number 0. See Figure 45:

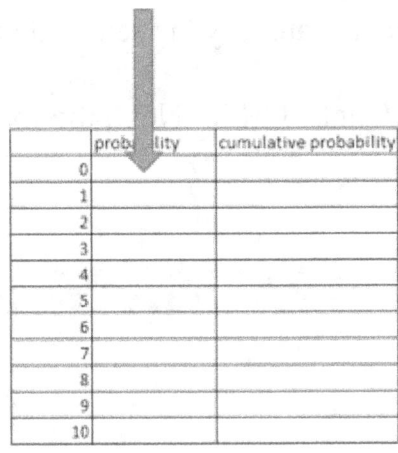

Figure 45: Binomial Setup Part II

Next select *FORMULA* and the *Insert Function* from the menu bar. See Figure 46.

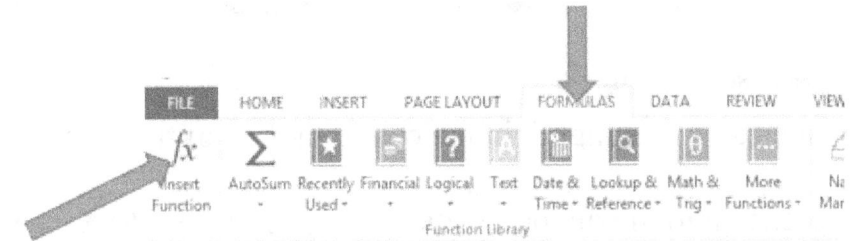

Figure 46: Binomial Setup Part II

The following dialogue box appear. See Figure 47:

Figure 47: Binomial Setup Part II

Make sure the "Statistical" category" is selected and scroll down until BINOM.DIST is displayed. Click "OK". Next, the following dialogue box appear. See Figure 48.

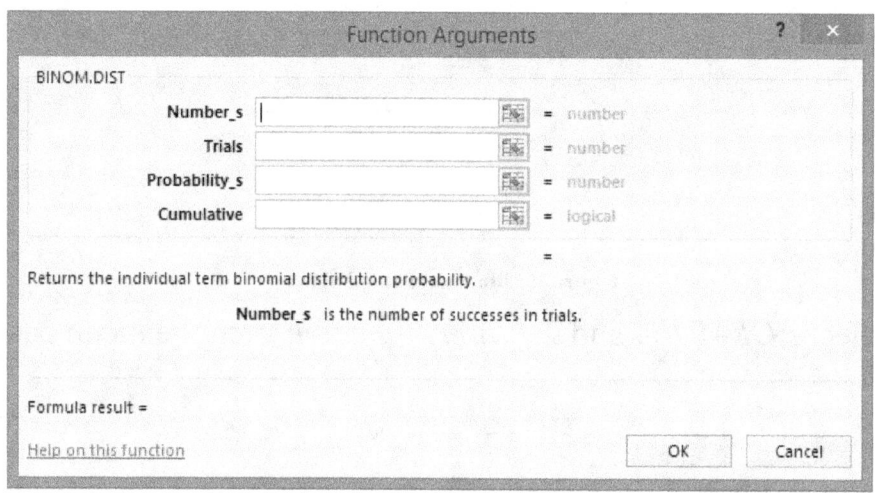

Figure 48: Binomial Setup Part III

In this example, one can either highlight the cell to the left of the cursor or simply type the cell reference into the dialogue box (in this example, F6), type in 10 for trials, .50 for probability and FALSE for cumulative. Before selecting OK, this is how the dialogue box will look like Figure 49.

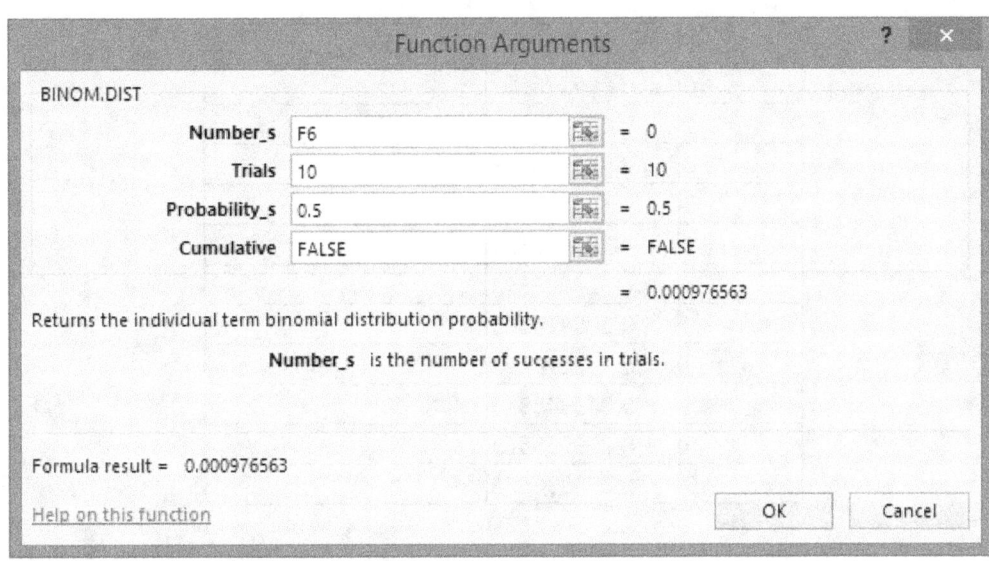

Figure 49: Binomial Setup Part IV

The result will look like Figure 50:

	probability	cumulative probability
0	0.000976563	
1		
2		
3		
4		
5		
6		
7		
8		
9		
10		

Figure 50: Binomial Setup Part V

Copy the first cell and paste into the remaining ten cells immediately underneath the first cell calculation. The result will look like Figure 51:

	probability	cumulative probability
0	0.000976563	
1	0.009765625	
2	0.043945313	
3	0.1171875	
4	0.205078125	
5	0.24609375	
6	0.205078125	
7	0.1171875	
8	0.043945313	
9	0.009765625	
10	0.000976563	

Figure 51: Binomial Setup Part VI

Please note several things. Compare the result generated next two the "2" cell and next to the "5" cell. These are exactly the same results as calculated performing the manual calculation. The calculations can also be double checked by adding all the probabilities up. If performed correctly, all the probabilities will add up to one. A quick way to do this is to highlight all the cells and use the auto sum function from the home tab. The current example is shown in Figure 52 with the sums of probabilities equaling zero.

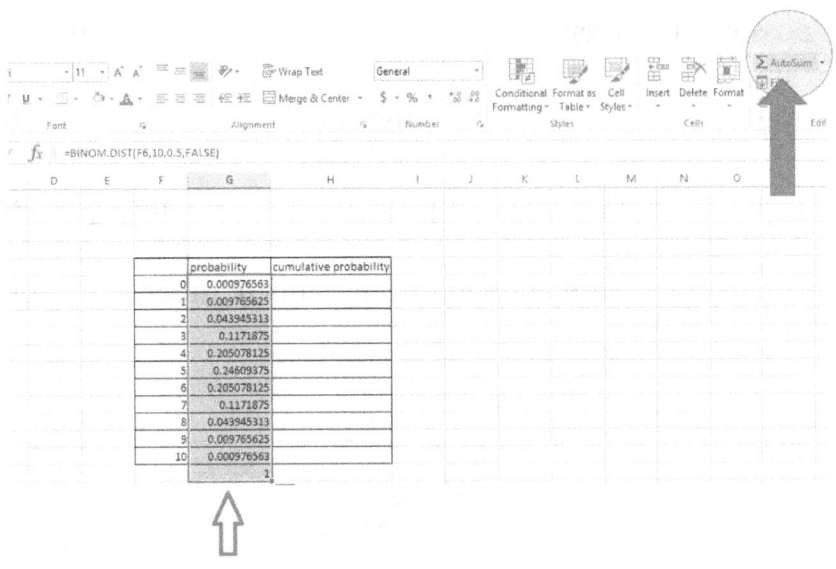

Figure 52: Binomial Setup Part VI

For the last column (cumulative probability), do exactly the same thing to get the cumulative distribution, with the exception of entering "TRUE" in the cumulative box (the last one in the BINOM.DIST dialogue box). The output will be look like the Figure 53. Notice the last cumulative probability will be equal to one. If this does not occur, the problem has not been solved correctly.

	probability	cumulative probability
0	0.000976563	0.000976563
1	0.009765625	0.010742188
2	0.043945313	0.0546875
3	0.1171875	0.171875
4	0.205078125	0.376953125
5	0.24609375	0.623046875
6	0.205078125	0.828125
7	0.1171875	0.9453125
8	0.043945313	0.989257813
9	0.009765625	0.999023438
10	0.000976563	1

Figure 53: Binomial Setup Part VII

It is also possible to determine the number of observations of each possible outcome by multiplying the probability by the number. In this case, the sum of the number of expected is 5. Recall that the probability of obtaining a heads is .50 and there were a total of 10 trials, so it is expected that this number is 5. See Figure 54.

	probability	# expected
0	0.000976563	0
1	0.009765625	0.009765625
2	0.043945313	0.087890625
3	0.1171875	0.3515625
4	0.205078125	0.8203125
5	0.24609375	1.23046875
6	0.205078125	1.23046875
7	0.1171875	0.8203125
8	0.043945313	0.3515625
9	0.009765625	0.087890625
10	0.000976563	0.009765625
	Total	5

Figure 54: Sum of Number of Expected Observations

We can also calculate the standard deviation of this binomial problem. The standard deviation is calculated by using the following equation:

$$\sigma = [nP(1-P)]^{1/2}$$

In this case, the standard deviation would be $[10*.5*.5]^{1/2} = 2.5^{1/2}$ or 1.58. One hint, something to the 0.5 power is same as taking square root of the number.

Chapter 6 Exercises:

1. Find P(X=1) given n=13 and p=.42

2. Find P(X=4) given n=13 and p=.42

3. Find P(X=12) given n=13 and p=.42

4. Find P(X=0) given n=8 and p=.33

5. Find P(X=4) given n=8 and p=.33

6. Find P(X=6) given n= n=8 and p=.33

7. Find P(X=10) given n=15 and p=.92

8. Find P(X=13) given n=15 and p=.92

9. Your top sales executive makes a sale 70% of the time she interacts with a customer. Given that there are 13 customers in the last hour of the day, what are the probabilities of making sales in the range of 0 to 13 customers? What is the expected standard deviation of sales?

10. Your administrative assistant calls in sick 35% of the time. Given there are 10 work days in the next two week period, probabilities of her/him missing days in the range of 0 to 13 days. What the expected standard deviation is of days missed.

11. A liver transplant is successful 90 percent of the time. You will perform 6 operations next month. Calculate the probabilities of failure from 0 to 6 failures and also calculate the cumulative failure rate.

Chapter 6 Solutions:

1. 0.0079

2. 0.1652

3. 0.0002

4. 0.0406

5. 0.1673

6. 0.0162

7. 0.0043

8. 0.2273

9.

	probability of sales		n	13
0	1.59432E-07		p	0.7
1	4.83611E-06		sd	1.652271
2	6.77056E-05			
3	0.000579259			
4	0.00337901			
5	0.014191843			
6	0.044152399			
7	0.103022265			
8	0.180288963			
9	0.233707915			
10	0.218127387			
11	0.138808337			
12	0.05398102			
13	0.009688901			
sum	1			

10.

	probability of missed days		n	10
0	0.013462743		p	0.35
1	0.072491695		sd	1.50831
2	0.175652953			
3	0.252219625			
4	0.237668493			
5	0.153570411			
6	0.0689098			
7	0.021203015			
8	0.004281378			
9	0.000512302			
10	2.75855E-05			
sum	1			

11.

	probability of failures	Cumulative
0	0.531441	0.531441
1	0.354294	0.885735
2	0.098415	0.98415
3	0.01458	0.99873
4	0.001215	0.999945
5	0.000054	0.999999
6	0.000001	1
Sum	1	

CHAPTER 7:   POISSON DISTRIBUTIONS

The second major discrete distribution that will be examined is the poisson distribution. The poisson distribution characterizes the probability of an event in an area of opportunity. In common practice, this is usually time, but could also be distance, area, length, volume, etc... For the poisson distribution to be applicable, several conditions must apply. First, the probability of an event at any given time is constant. In other words, the average rate remains the same. Second, the occurrence of an event must be independent of previous events. In other words, the occurrence of one event does not influence the occurrence, or lack of occurrence, of another event. Finally, the probability that two or more events will occur at the same time, or other measurements, gets smaller as the area gets smaller. For example, the probability of getting stung twice by a bee is higher in a month long time period than in a 10 minute time period.

The formula that is used to calculate the poisson distribution is as follows:

$$f(x) = [(nP)^x \, e^{-nP}] / x!$$

x = observation of interest
e = base of natural logarithms (2.71828)
P = known probability
n = number given of total population
np = may be given as mean

Using this formula, the probability of x occurring during/within the stated measurement can be calculated. Before examining how Microsoft® Excel help with these calculations, it is instructive to solve a simple problem. If the average employee takes two breaks per eight hour shift, what is the probability that an employee will take four breaks per shift? This problem is solved in the following manner:

\# of breaks = **[(nP)^x e^{-np}] / x!** = **[(2)^4 *2.71828^{-2}] /4!** = **(16*0.135335)/24 = 0.09223**

Microsoft® Excel makes poisson problems easy. One can solve individual probabilities or one can solve multiple probabilities. In Microsoft® Excel, the first step is to open "insert function" from the "formula" heading on the menu bar. See Figure 55:

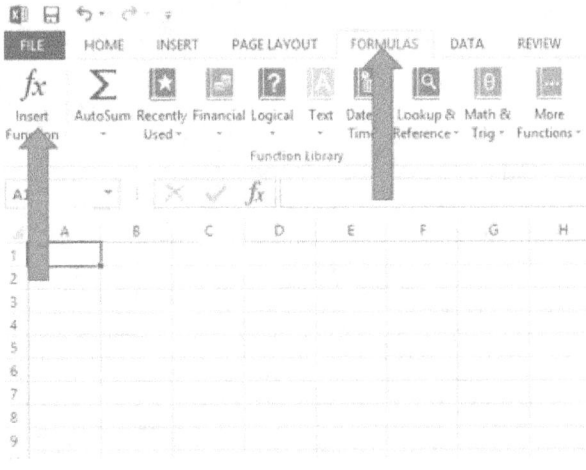

Figure 55: Poisson Setup I

Next, from the "statistical" category, select POISSON.DIST. See Figure 56:

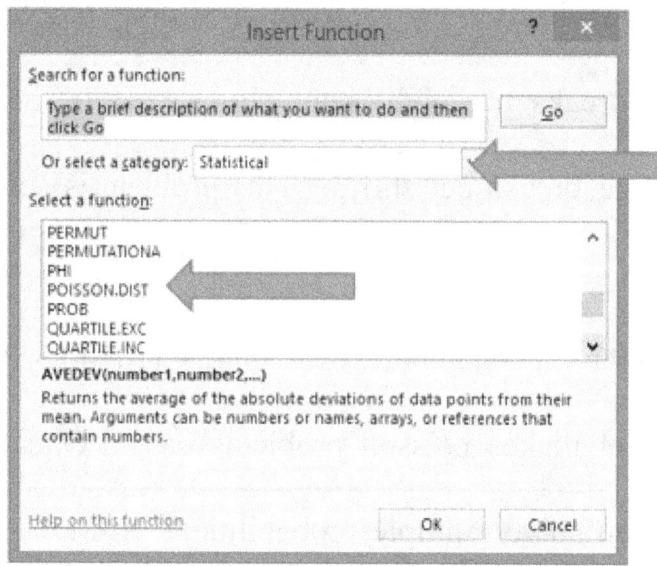

Figure 56: Poisson Setup II

Next the following dialogue box will be seen: See Figure 57:

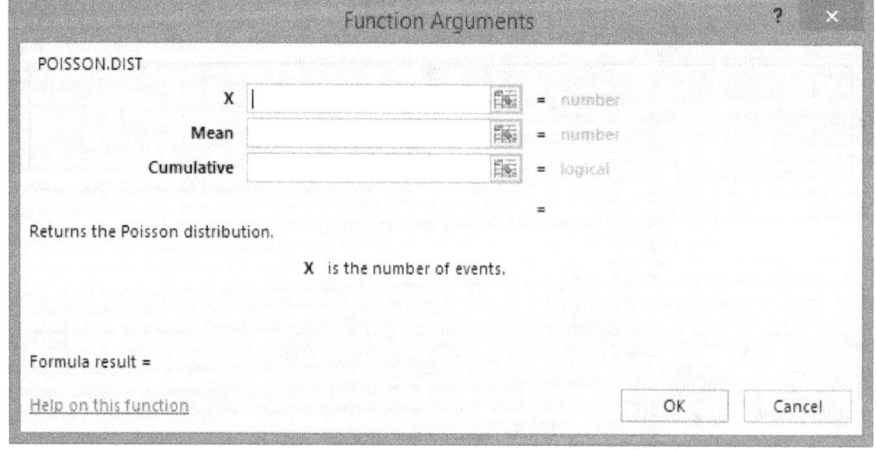

Figure 57: Poisson Setup III

In the "X" box, put the probabilty of interest. In the previous example, this would be 4. In the "Mean" box, put the given or calculated mean. In the previous example, this would be 2. If looking for a single answer, put "false" in cumulative. Press OK, and the answer will appear. Figure 58 is an how the example answer would look. Notice that the answer is exactly the same as the way we calcualted it.

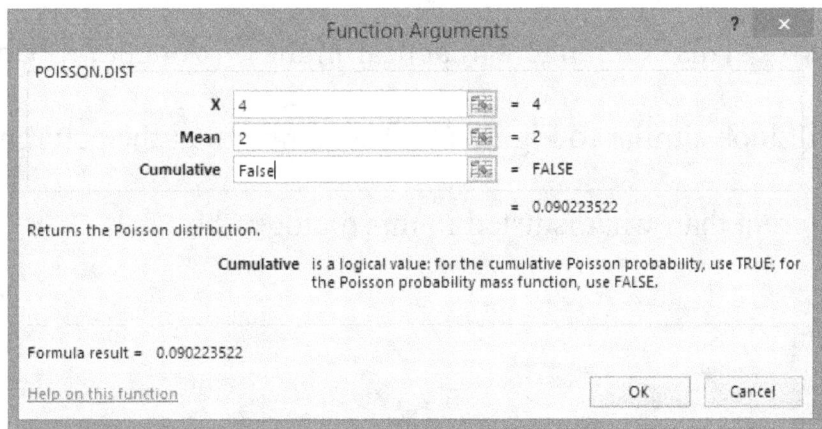

Figure 58: Poisson Example Output

Using the previous example, we can also use excel to answer not only how many times would expect four breaks, but we can also quickly calculate how many of all breaks would be taken. The first step is to set up the data and place the cursor in the first cell under probability. See Figure 59.

Breaks	Probability
0	
1	
2	
3	
4	
5	
6	
7	

Figure 59: Poisson Data

Next, open the POISSON.DIST from "insert function". Highlight the cells that contains 0 to 7. This cell range will appear in the box under probability. The dialogue box will look similar to Figure 60. Remember that the cell range will probably be different than what is listed in this dialogue box.

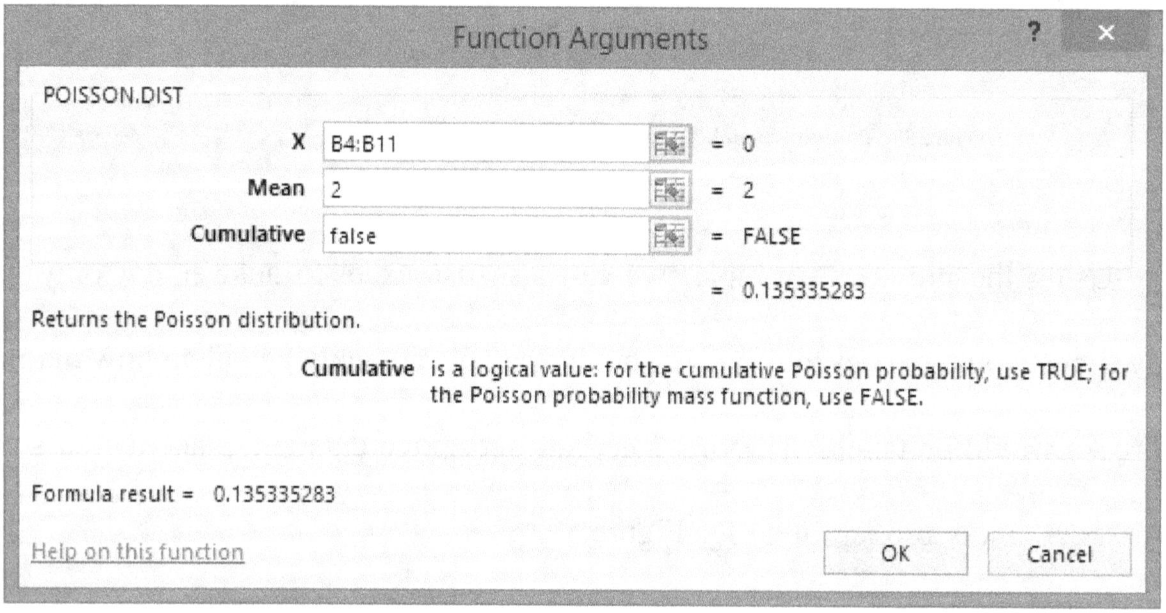

Figure 60: Poisson Dialogue Box

From this dialogue box, it is determined that the probability of zero breaks is 0.1353. The data in the workbook will appear as follows in Figure 61:

Breaks	Probability
0	0.135335283
1	
2	
3	
4	
5	
6	
7	

Figure 61: Poisson Probability of Zero

Next, copy and paste the value in the first cell under "Probability" and cut and paste into the next seven cells. The output should look like Figure 62:

Breaks	Probability
0	0.135335283
1	0.270670566
2	0.270670566
3	0.180447044
4	0.090223522
5	0.036089409
6	0.012029803
7	0.003437087

Figure 62: Poisson Probability of Zero II

One may wonder what percentage of the breaks this accounts for. There are two ways to do this. A separate column could be created and a similar process followed with the exception that "true" is placed in the cumulative box. A quicker way is to simply sum the probabilities. The percentage of total breaks taken is 99.89%. The percentage of employees taking more than seven breaks is 0.11%. This is summarized in Figure 63:

Breaks	Probability
0	0.135335283
1	0.270670566
2	0.270670566
3	0.180447044
4	0.090223522
5	0.036089409
6	0.012029803
7	0.003437087
	0.998903281

Figure 63: Poisson Total Break Summary

Chapter 7 Problems

1. An average of ten drivers an hour are seen texting while driving through an intersection. What is the probability that you will observe six drivers texting in the next hour?

2. An average of 100 newspapers are sold in a given day. What are the probabilities that 95 newspapers will be sold? What are the probabilities that 101 will be sold?

3. An average of six players win a game at a carnival during the last hour of operations. What are the probabilities of zero to ten winners of the game?

4. Calculate the cumulative probabilities from problem #3.

Chapter 7 Solutions

1. 0.0630

```
POISSON.DIST
    X            6                = 6
    Mean         10               = 10
    Cumulative   false            = FALSE
                                  = 0.063055458

Returns the Poisson distribution.
    Cumulative is a logical value: for the cumulative Poisson probability, use TRUE; for
    the Poisson probability mass function, use FALSE.

Formula result = 0.063055458
```

2. 3.6%, 3.9%

POISSON.DIST

X	95	= 95
Mean	100	= 100
Cumulative	FALSE	= FALSE
		= 0.036012427

Returns the Poisson distribution.

Mean is the expected numeric value, a positive number.

Formula result = 0.036012427

POISSON.DIST

X	101	= 101
Mean	100	= 100
Cumulative	FALSE	= FALSE
		= 0.039466333

Returns the Poisson distribution.

X is the number of events.

Formula result = 0.039466333

3.

WINNERS	PROBABILITY
0	0.002478752
1	0.014872513
2	0.044617539
3	0.089235078
4	0.133852618
5	0.160623141
6	0.160623141
7	0.137676978
8	0.103257734
9	0.068838489
10	0.041303093

4. 95.73%

WINNERS	PROBABILITY	CUMULATIVE
0	0.002478752	0.002478752
1	0.014872513	0.017351265
2	0.044617539	0.061968804
3	0.089235078	0.151203883
4	0.133852618	0.2850565
5	0.160623141	0.445679641
6	0.160623141	0.606302782
7	0.137676978	0.74397976
8	0.103257734	0.847237494
9	0.068838489	0.916075983
10	0.041303093	0.957379076

CHAPTER 8: INTRODUCTION TO INFERENTIAL STATISTICS

Multiple textbooks could be written about just one of the many topics of inferential statistics. It is certainly not the intent of this very short introduction to propose to be adequate. The reader should defer to other sources for a more complete description of the "nuts and bolts" of inferential statistics. However, Microsoft® Excel does a great job of providing us the tools to become a frequent user of inferential statistics.

The mere description of data allows us to learn more and make some judgments on what we think may be the case. For example, one may take repeated measurements of the temperature in North Pole Alaska in January and make a judgment, based on observations of the data, that it is "cold". One could also measure the hay consumption of llamas in Preston Idaho and make the observation that llamas eat hay in Preston Idaho. In most cases, the scientific or personal question being asked requires more than just a quick judgment on the presented data. What if the question was "is it colder in January or February in North Pole Alaska?" What if the question was "do llamas or cows eat more hay per day in Preston Idaho?" Inferential statistics help us make intelligent judgments based on data.

There are some very basic things that must be considered before continuing in this chapter. First, there is an important distinction that must be made between the terms "statistic" and "parameter". A statistic is calculated from a sample. A parameter is a characteristic of a population. After a statistic is calculated from a sample, inferences can be made about the population.

A researcher will usually begin with an idea or thought about what might be occurring in a given situation. This is called their hypothesis. The null hypothesis is just the opposite and supposes that the hypothesis is not occurring. In the research world, the opposite of the null hypothesis is called the alternate hypothesis. In practice, rejection of the null hypothesis indicates the evidence supports the belief the null hypothesis is unlikely true. This does not mean that the null hypothesis must not be true, rather, the evidence strongly supports it is not true. In other words, rejection of the null hypothesis supports the case that the hypothesis is "true". The researcher will set the parameters by which these judgments are made.

No matter what test is selected, the researcher will compare the test statistic to defined areas of rejection and nonrejection. If the test statistic is in the nonrejection area, then the null hypothesis can't be rejected. Therefore, it is very important to know the line of debarkation between the area of rejection and nonrejection. This line is known as the critical value. See Figure 64.

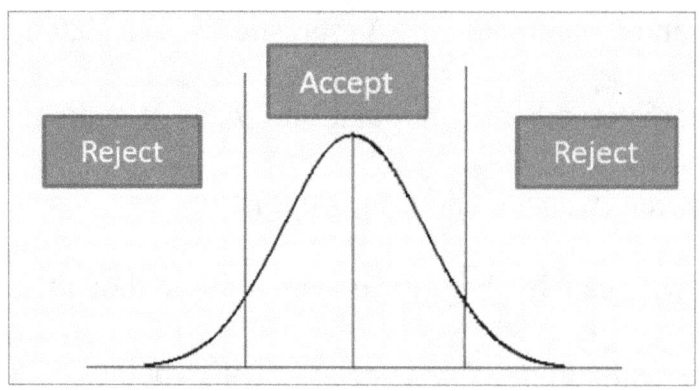

Figure 64: Areas of Rejection and Acceptance

With any judgment of a hypothesis test, there are real risks of making errors. The following is a summary of the types of errors that can be made:

Your Decision	Assertion Is True	Assertion Is False
Accept	CORRECT	TYPE II ERROR
Reject	TYPE I ERROR	CORRECT

- Type I Error = False Positive: rate of Type I = Alpha
- Type II Error = False Negative rate of Type II = Beta
- Power of Test = 1-Beta

Figure 65: Errors Associated With Inferential Statistics

The probability that the researcher will commit a type I error (α) is also known as the level of significance. The most common α is 0.05 or a one in twenty chance of committing a type I error.

CHAPTER 9: NORMAL DISTRIBUTIONS

Data can be distributed in many different ways. This text will cover four distinct distributions. The covered distributions are the F, t, chi-squared, and normal distribution. The later will be discussed in this chapter. The understanding of the normal distribution of data is an important concept for the student of statistics as it has enormous implications in inferential statistics. A very simplistic description of the normal distribution include these four general properties.

- The mean, median, and mode are the same

- The data is curve shaped (bell-shaped) and there is symmetry about the mean.

- The total area under the curve(AUC) is one

- The normal curve approaches the x-axis, but never touches the x-axis as it extends away from the mean in both direction

One common misconception is that a "normal curve" looks the same, no matter what the data is composed of. This is not the case and is demonstrated by Figure 66:

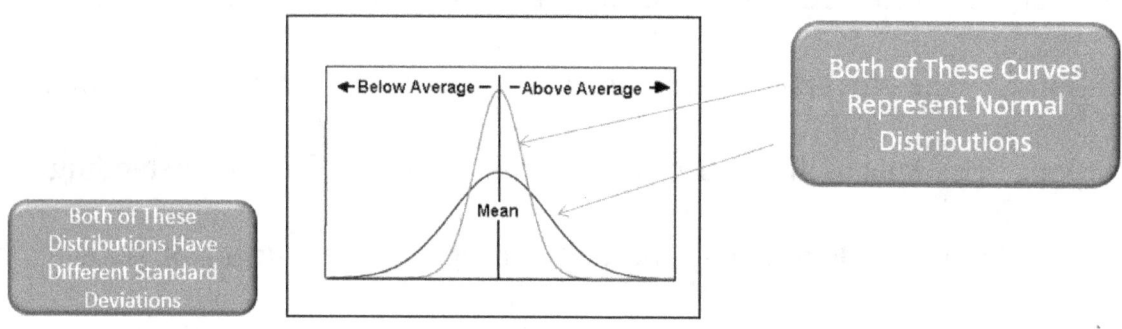

Figure 66: Shape of Normal Distribution

To fully understand and characterize the normally distributed curve, one must have two pieces of data. The first is mean and the second is the standard deviation. With this information, one can predict the area under the curve and determine the probability of any particular observation occurring there. Conversely, given a data point, one could determine what percentage of data points would occur to the right or to the left of the data point. The percentage of the area under the curve for plus/minus three standard deviations is shown in Figure 67:

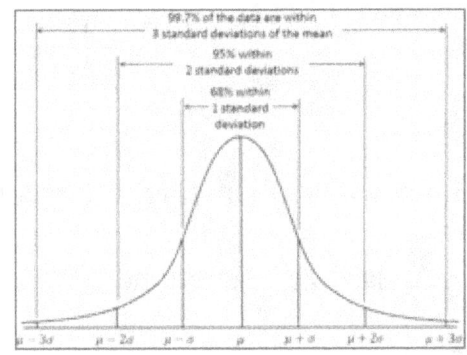

Figure 67: Mean ± 3 SD

The normal standard curve is defined by a mean value of zero (e.g. $\mu = 0$) and a standard deviation of one (e.g. $\sigma = 1$). To transform a datapoint into a useful number, we calculate the standard normal deviate (Z). Z is calculated by subtracting the datapoint of interest from the mean and dividing by the standard deviation. This is shown in the following formula.

$$Z = (X_i - \mu)/\sigma$$

For example, suppose a task takes an employee a mean 90 minutes to complete and has a standard deviation of 10 minutes. A new employee takes 110 minutes. The Z score for this is $(110-90)/10 = 2.0$. This suggests the new employee completes the task two standard deviations slower than the average employee. If the employee had completed the task in 70 minutes, the Z score for this is $(70-90)/10 = -2.0$ and suggest the employee completed the task faster than the average employee. Notice that the Z score is the same with the exception of the sign (+/-). Examine Figure 67 and notice the relative distance between points on the normal curve. Notice that 95%

of the area under the curve is between -2 standard deviations and +2 standard deviations.

Microsoft® Excel help with calculating a Z score. In Microsoft® Excel, it is calculated using the "standardize" function. To use this function, select "insert function" from "formulas" on the menu bar. Alternatively, simply hit "Shift + F3". See Figure 68:

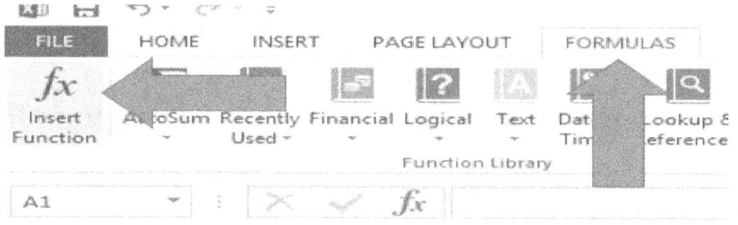

Figure 68: Z Score Step 1

From there, a dialogue box will appear and "standardize" should be selected. See Figure 69:

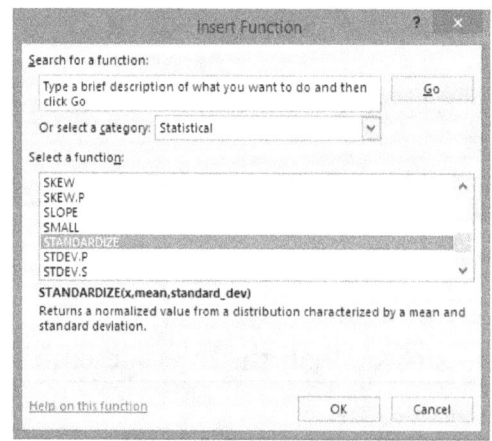

Figure 69: Z Score Step 2

After selecting "OK", another dialogue box will appear. See Figure 70:

Figure 70: Z Score Step 3

The same items needed to manually solve for a Z score will be prompted in this dialogue box; the item to be Z scored (X), the mean, and the standard deviation. For example, if the need is to calculate a Z score for 10, with a mean of 8, and and a standard deviation of .75, the result will be 2.66. The dialouge box showing the setup and result follows. See Figure 71"

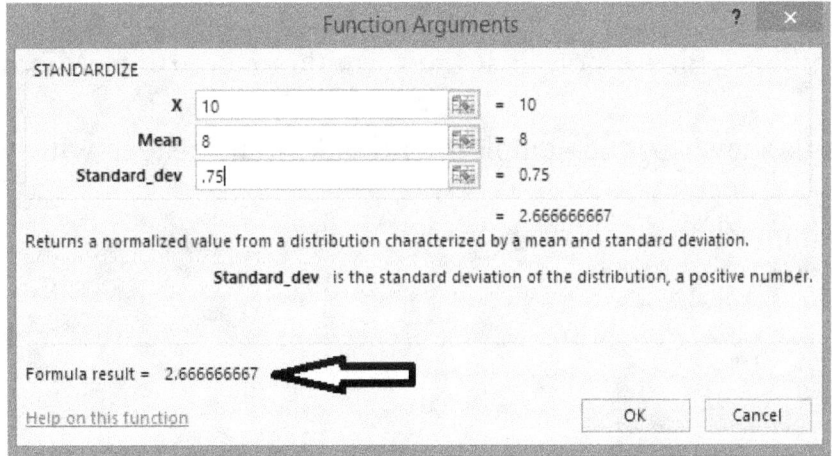

Figure 71: Z Score Step 4

It is also possible to transform a given Z score to determine either the point of interest, the standard deviation, or mean. This is done by performing simple algebra by rearranging the equation to solve for the unknown. For example, if given a Z score of 2.0, a mean of 100, and a standard deviation of 50, the unknown data point can be determined by using the following:

$$2.0 = (x-100)/50$$

$$2.0*50 = x-100$$

$$100+100 = x$$

$$x = 200$$

Using simple Microsoft® Excel formulas, these unknowns can be calculated. This will take a little work to input the formulas, and it may be quicker to do the math manually, but this is a great way to check the work. The following are examples of how to use simple Microsoft® Excel formulas to solve for Z, the data point, the mean, and the standard deviation. For these types of problems, the unknown is solved using three given numbers. Please note that the formula entered into Microsoft® Excel is based on what is shown in the sample workbook. The reader will have to modify their formulas to match the cells of interest.

Example of Formula to Solve for Z (Figure 72):

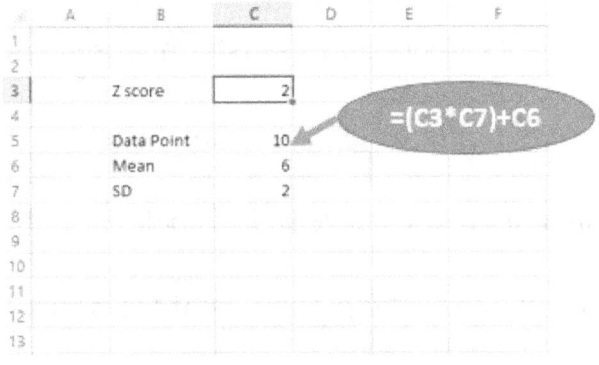

Figure 72: Z Score Calculation Example I

Example of Formula to Solve for Data Point (Figure 73):

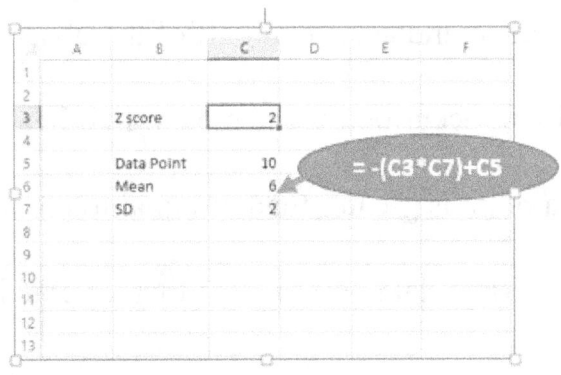

Figure 73: Z Score Calculation Example II

Example of Formula to Solve for Mean (Figure 74):

Figure 74: Z Score Calculation Example III

Example of Formula to Solve for Standard Deviation. Note that "ABS" precedes the formula. This is because the absolute value is needed to ensure the standard deviation will not be negative.

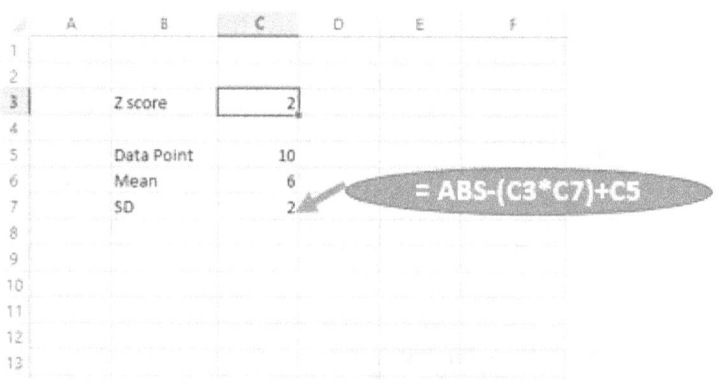

Figure 75: Z Score Calculation Example IV

A Z score alone may not be enough to satisfy the researcher. Instead of just knowing what the Z score, it might be good to know what the probability of a future event occurring in the area under the curve (AUC) below or above the particular Z score. Keeping in mind that one of the properties of the normal distribution is the symmetry about the mean, probabilities are easily calculated. The first thing to consider is a particular area that confuses many students. It must be remembered that the symmetry around the mean indicates that half of the observations will occur above the mean and half will occur below the mean. In other words, the probability that any one random data point is greater than or equal to the mean is 0.500 and the probability that any one random data point is less than or equal to the mean is 0.500.

Keeping this one observation in mind will make solving problems much easier. See Figure 76:

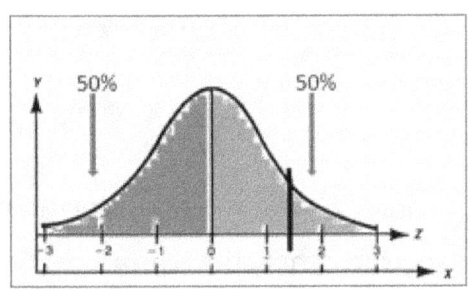

Figure 76: Area Under Curve > and < Than Mean

To solve problems involving areas under the curve and probabilities associated with the area under the curve, there are two methods. The first is manual calculation and the second is to use electronic means such as Microsoft® Excel. Both methods are easy and straight forward. To fully understand how Microsoft® Excel solves this these types of problems, it is best to understand how to solve these problems manually. Refer to Table I in the back of the text. This is a table of the Standard Normal Distribution. Using Table I and given a Z score of 2.5, one can easily look up the tabular value which is 0.49379. To calculate the area under the curve that is greater than a Z of 2.5, simply subtract 0.49379 from 0.50000. The result is 0.00621. This is easily visualized by estimating where this data point would be on the normal curve. There is not much room under the curve to the right of Z=2.5. See Figure 77:

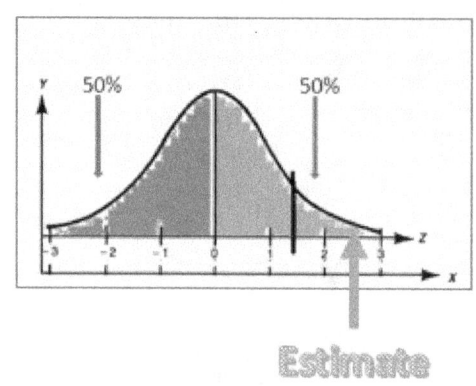

Figure 77: Small Area To Right Of Data Point

The reverse is also true when determining the area under the curve less than Z=2.5. In this case, calculate this probability by adding 0.5000 (the area under the curve less than zero) to 0.49379. The resulting area under the curve is 0.99379. In other words, more than 99% of the area under curve exists less than Z=2.5. In using the Table I, there may be a need to look up a tabular value for a number that has not been rounded. For example, what if the number of interest was Z=2.51. Refer back to Table I and notice that to the right of Z at the top of the table are decimals to the hundredth place. The tabular value for 2.51 would be 0.49396. In the event the number of interest is to the thousandth place, simply add the two numbers together and divide by 2. For example, to get the tabular value for 2.505, simply add the tabular value of 2.50 and 2.51 together and divide by 2. One of the major advantages of using Microsoft® Excel is not needing to worry about referring to a Table of

Standardized Normal Distributions. Refer to Figure 78 to examine how the NORMDIST formula can give the same result as obtained from Table I.

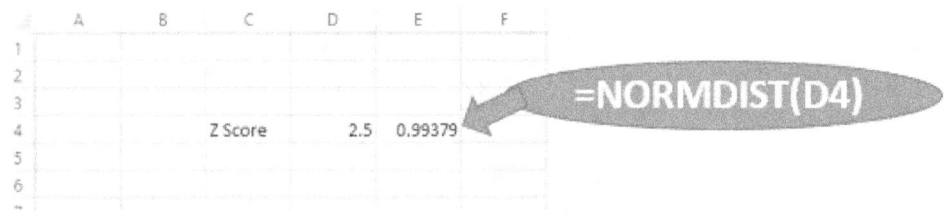

Figure 78: NORMDIST Formula

There are many times when the area under the normal curve is desired between two Z scores. For example, one may be interested in determining the probability of a data point falling under the curve between two data points. For example, given a Z scores of 0.5 and 2.0, what is the probability of a data point falling between the two Z scores? Referring to Table I will give a value of 0.1915 for a Z score of 0.5 and 0.4722 for a Z score of 2.0. The difference between the two scores is 0.4722-0.1915 which is 0.2857 which means there is a 28.57% probability that the data point would fall between the two Z scores. This method can be used to find the area between two Z scores on a curve with a normal distribution. Microsoft® Excel can also be used to quickly determine the area under the curve between two Z scores. For example, using the same general excel formulas for Z scores, one can subtract the difference to find the probability for the area under the curve. In the following example, the area between Z=1 and Z=-1 have been calculated (68.2%) using Microsoft® Excel. See Figure 79:

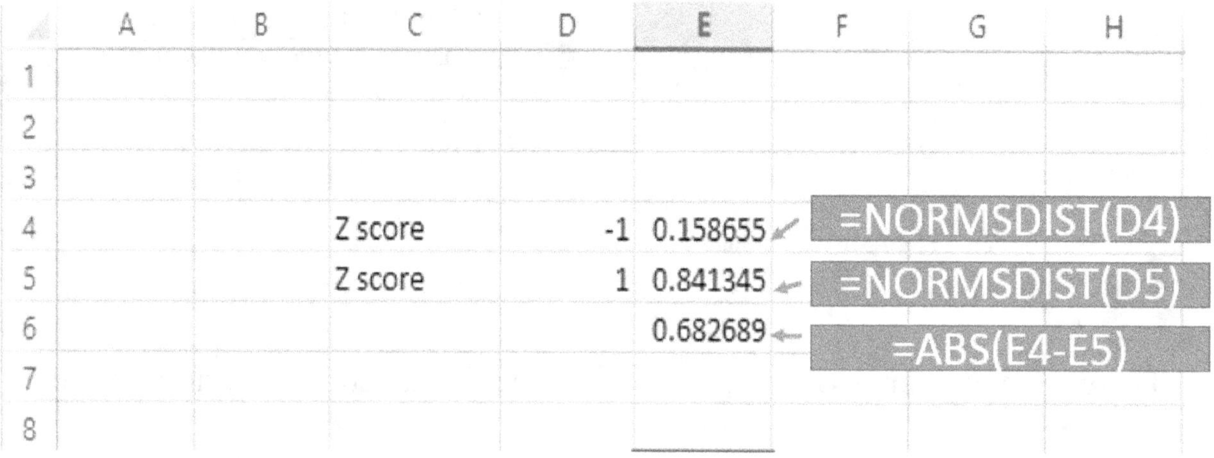

	A	B	C	D	E	F	G	H
1								
2								
3								
4			Z score	-1	0.158655	=NORMSDIST(D4)		
5			Z score	1	0.841345	=NORMSDIST(D5)		
6					0.682689	=ABS(E4-E5)		
7								
8								

Figure 79: Determination of Probability of Data Point Occurring Between 2 Z Scores

Chapter 9: Problems

1. Calculate the Z score of 15 where the mean is 10 and the standard deviation is 4.

2. Calculate the Z score of 10 where the mean is 15 and the standard deviation is 4.

3. Calculate the Z score of 6 where the mean is 4 and the standard deviation is 2.

4. Calculate the Z score of 150 where the mean is 180 and the standard deviation is 30.

5. Calculate the Z score of 110 where the mean is 109 and the standard deviation is 10.

6. Calculate the Z score of 17 where the mean is 13 and the standard deviation is 5.

7. Calculate the Z score of 150 where the mean is 100 and the standard deviation is 40.

8. Calculate the Z score of 1500 where the mean is 1400 and standard deviation is 144.

9. Calculate the Z score of 3 where the mean is 2 and the standard deviation is 0.5.

10. Calculate the Z score of 100 where the mean is 90 and the standard deviation is 40.

11. Calculate the mean given the Z score is 1, the datapoint is 20, and the standard deviation is 10.

12. Calculate the mean given the Z score is -1, the datapoint is 20, and the standard deviation is 10.

13. Calculate the mean given the Z score is 4, the datapoint is 50, and the standard deviation is 20.

14. Calculate the datapoint given the Z score is 2, the mean is 15, and the standard deviation is 8.

15. Calculate the datapoint given the Z score is 3, the mean is 30, and the standard deviation is 5.

16. Calculate the datapoint given the Z score is 1, the mean is 15, and the standard deviation is 12.

17. Calculate the datapoint given the Z score is -0.5, the mean is 10, and the standard deviation is 3.

18. Calculate the standard deviation given the Z score is -0.5, the mean is 10, and the datapoint is 8.

19. Calculate the standard deviation given the Z score is 0.5, the mean is 20, and the datapoint is 18.

20. Calculate the probability that that a datapoint will fall between Z=-3 and Z=-1.

21. Calculate the probability that that a datapoint will fall between Z=-3 and Z=1.

22. Calculate the probability that that a datapoint will fall between Z=0 and Z= 3.

23. Calculate the probability that that a datapoint will fall between Z=-1 & Z= 2.

Chapter 9: Solutions

1. 1.25

2. -1.25

3. 1.00

4. -1.00

5. 0.1

6. 0.8

7. 1.25

8. 0.69

9. 2.00

10. 0.25

11. 10

12. 30

13. -30

14. -1

15. 15

16. 3

17. 11.5

18. -4

19. -366

20. 0.1573

21. 0.8399

22. 0.4987

23. 0.8186

CHAPTER 10: CONFIDENCE INTERVALS

There are two major types of estimates used in statistical inference. The first is point estimates. Point estimates are single sample statistics used to estimate a particular population parameter. For example, when the mean is derived from a sample, it is in fact a point estimate of the population mean. It is known that point estimates have inherent variability and the "truth" lies somewhere around the point estimate. It is also known that the larger the sample, the better chance the point estimate accurately reflects the population parameter. In other words, the larger the sample, the less "spread" there is around the point estimate. Interval estimates can be used to estimate the spread around the population parameter of interest. These interval estimates will be known as "confidence intervals" in this chapter.

It is important to note the sample size when considering confidence intervals. Larger samples (n >30) conform to the central limit theorem. The central limit theorem suggests that larger samples will be normally distributed. Many of the same principles used during "Z scoring" are used in developing confidence intervals with large samples. In practice, the most common confidence interval is 95%. The 95% confidence interval is bi-directional, meaning that the remaining 5% is found in both directions, or 2.5% in each direction. From Table I, the corresponding value for $(100-\alpha)/2$ is 0.475. After finding 0.475 in Table I, the normal deviate is found to be ±1.96. The formula for the confidence interval is shown in Figure 80:

$$C.I. = \bar{X} \pm Z_{\frac{\alpha}{2}} \frac{\sigma}{\sqrt{n}}$$

Figure 80: Confidence Interval

Consider a sample of 100 employees from a given population. The mean employee evaluation rating during their annual performance evaluation was "nine" with a standard deviation of "one". The 95% confidence interval would be: 9 ± 1.96(1/10) or 8.804 to 9.196. Microsoft® Excel easily calculate confidence intervals on large samples. Consider the following sample of 36 weights of meat (Figure 81):

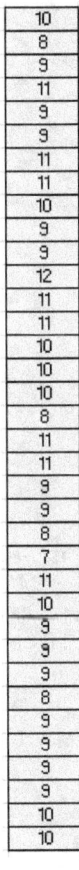

Figure 81: Weights of Meat

The first step in developing a confidence interval is to determine the mean and standard deviation. This is easily done in excel using the data analysis function. See Figure 82:

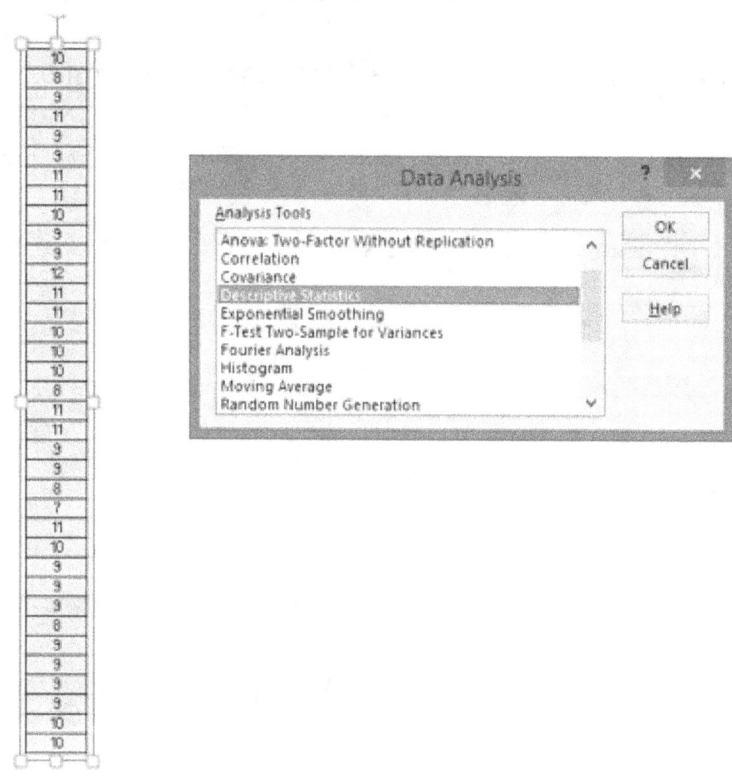

Figure 82: Descriptive Stats of Weights of Meat

The output will appear as follows in Figure 83:

Column1	
Mean	9.583333
Standard Error	0.188457
Median	9
Mode	9
Standard Deviation	1.130739
Sample Variance	1.278571
Kurtosis	-0.41477
Skewness	0.03226
Range	5
Minimum	7
Maximum	12
Sum	345
Count	36
Confidence Level(95.0%)	0.382587

Figure 83: Descriptive Stats - Weight of Meat Output

Similar to the previous problem concerning employee evaluations, the confidence interval can be easily calculated.

$$C.I. = \overline{X} \pm Z_{\frac{\alpha}{2}} \frac{\sigma}{\sqrt{n}}$$

9.58 ± 1.96(1.13/6) or 9.58 ± 0.369 or 9.211 to 9.949

Instead of inputting a new formula every time a new confidence interval is desired, consider using the "confidence" function in excel. To use the confidence function, simply type "=confidence.norm (alpha, std dev, sample size) into the cell. For example, in our previous example of weights of meat, the following would by typed into the cell "=confidence.norm(.05,1.13,36)" and the result will be 0.3691, the exact same result from our manual calculation. Please note that this does not apply to samples with less than 30 observations as this confidence interval applies to "normal distributions". For samples with less than 30 observations, the "confidence.t" function should be used.

Chapter 10 problems: Calculate 95% confidence interval

1. If the mean is 100, the standard deviation is 8, and the sample size is 50, calculate the confidence interval for the population mean.

2. If the mean is 100, the standard deviation is 12, and the sample size is 50, calculate the confidence interval for the population mean.

3. If the mean is 50, the standard deviation is 8, and the sample size is 90, calculate the confidence interval for the population mean.

4. If the mean is 100, the standard deviation is 2, and the sample size is 150, calculate the confidence interval for the population mean.

5. If the mean is 100, the standard deviation is 8, and the sample size is 150, calculate the confidence interval for the population mean.

6. If the mean is 50, the standard deviation is 18, and the sample size is 50, calculate the confidence interval for the population mean.

7. If the mean is 1000, the standard deviation is 18, and the sample size is 350, calculate the confidence interval for the population mean.

8. If the mean is 10, the standard deviation is 1, and the sample size is 150, calculate the confidence interval for the population mean.

9. If the mean is 100, the standard deviation is 0.8, and the sample size is 40, calculate the confidence interval for the population mean.

10. If the mean is 25, the standard deviation is 8, and the sample size is 40, calculate the confidence interval for the population mean.

Chapter 10 solutions

1. 97.7 – 102.2

2. 96.7 – 103.3

3. 48.7 – 51.3

4. 99.7 – 100.3

5. 98.7 – 101.3

6. 45.0 – 55.0

7. 998.1 – 1001.9

8. 9.8 – 10.2

9. 99.8 – 100.2

10. 22.5 – 27.5

CHAPTER 11: t DISTRIBUTION

There was a reference to the t distribution in the previous chapter. t distributions are similar in appearance to normal distributions and share similar characteristics such as symmetry around the mean and bell shaped appearance. The most obvious difference between the t distribution and normal distribution is an elongated area under the tails of the curve in both directions. Intuitively, this makes sense as there is more uncertainty with smaller sample sizes. Recall, that t distributions are appropriate when sample sizes are less than 30. The closer the sample size is to 30, the closer the t distribution curve will resemble a normal distribution. This fact is important, as each t distribution has a slightly different curve. The t distribution is based on samples drawn from a population versus the normal distribution which is based on the entire population. The following figures (Figures 84 and 85, source Wikipedia) highlight the difference of the normal distribution compared to a sample t distribution with one and two degrees of freedom.

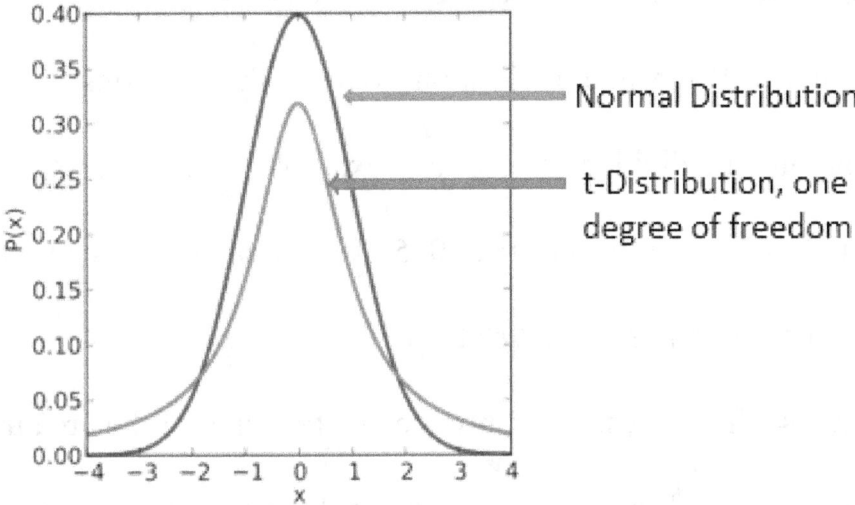

Figure 84: Normal vs. t-Distribution

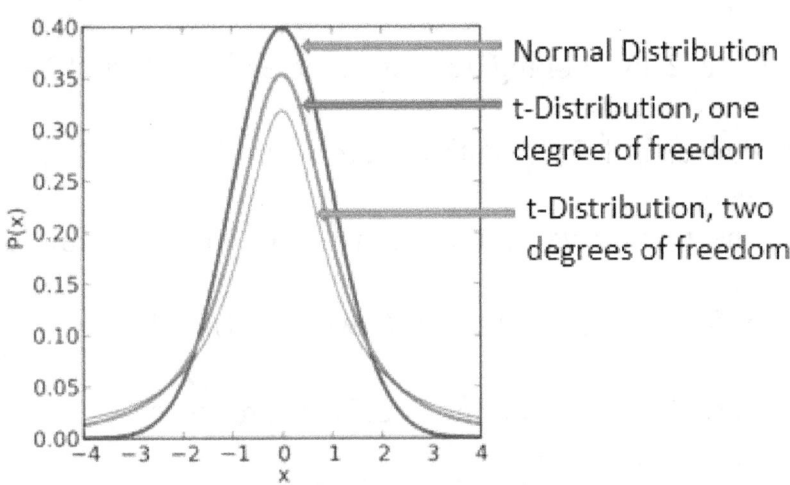

Figure 85: Difference in Appearance Between t-distributions

The degrees of freedom associated with the t-distribution is defined as n-1. For example, if there were 25 observations, then n-1 is equal to 24. The Z scores are calculated the same way as in normally distributed populations with the exception that a different data source is used (table 2).

It is important to note the difference between one and two tailed tests. Each test performed will have an alpha. In many cases, alpha will be 0.05. If the interest is in only one direction of the curve, then the area under that particular direction will be equal to alpha. If the interest in alpha is in both directions, positive and negative, then alpha must be divided by two. For example, if the interest is in both directions and alpha is 0.05, this indicates that .025 of the area under the curve is located in each tail of the distribution of the test statistic.

Suppose the interest is in only the positive side of the area under the tail of a t-distribution and the desire is to account for the most positive five percent of the area under the curve. Assume there are 25 observations. The tabular value for t can be found from table 2 by looking for the intersection of "n-1" or 24 and 95 % (one sided). The answer is 1.711. If the interest was in both directions with alpha being five percent, the tabular value for t can be found from table 2 by looking for the intersection of "n-1" or 24 and 95 % (two sided). The answer is 2.064. This area would encompass the upper and lower 2.5% of the curve.

Confidence intervals can be calculated manually for the t-distribution. The following formula can be used. See Figure 86.

$$\text{Confidence Interval (t distribution)} = X \pm t_{n-1}*(sd/\sqrt{n})$$

Figure 86: Confidence Interval for t-distribution

For example, if a 95% confidence interval is required for measurements of lengths of mittens, assume the following: 25 mittens, average length is 7.0 inches, and the standard deviation is 1.5 inches. From table 2, the tabular value for t, 0.025, 24 is 2.064. The confidence interval is calculated as follows:

95% Confidence Interval = 7.0 ± 2.064(1.5/5) or 6.38 to 7.62 inches

Confidence intervals can be calculated from the t distribution using the "=confidence.t" function of Microsoft® Excel. Using the same example, type "=confidence.t" into the cell. See Figure 87:

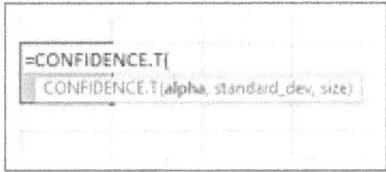

Figure 87: Initial CONFIDENCE.T

Next, the alpha, standard deviation, and sample size should be entered. It should look as follows – "=CONFIDENCE.T (0.05, 1.5, 25)". Please note than one does not need to divide the .05 into two separate halves. In other words, input the 0.05 and not 0.025 for alpha.

Problems Chapter 11:

1. Find the tabular t value for alpha .05 (1 way) and n=25

2. Find the tabular t value for alpha .05 (2 way) and n=15

3. Find the tabular t value for alpha .05 (1 way) and n=15

4. Find the tabular t value for alpha .05 (2 way) and n=25

5. Find the tabular t value for alpha .05 (1 way) and n=10

6. Determine the 95% confidence interval for a sample with n=29, sd =1, and mean =10

7. Determine the 95% confidence interval for a sample with n=25, sd =1, and mean =20

8. Determine the 95% confidence interval for a sample with n=20, sd =3, and mean =30

9. Determine the 95% confidence interval for a sample with n=19, sd =2, and mean =20

10. Determine the 95% confidence interval for a sample with n=9, sd =1, and mean =10

Solutions Chapter 11:

1. 1.7081
2. 2.1448
3. 1.7613
4. 2.0639
5. 1.8125
6. 9.28 – 10.72
7. 19.59-20.41
8. 28.6 – 31.4
9. 19.04 – 20.96
10. 9.33 – 10.77

CHAPTER 12: TESTING FOR SIGNIFICANT DIFFERENCES – LARGE SAMPLES

There are times when it is necessary to determine if there are differences between measured values and the population of interest. There are distinct differences in how this is accomplished for small sample sizes (<30) and large sample sizes (>30). Chapter 12 will cover large samples.

The first step in determining if there are differences between observed measurements and the true population mean is to know the number of samples. Next, a null hypothesis if formulated, which by definition, means that an alternate hypothesis will be generated. The alternate hypothesis can be one way (greater than or less than) or it can be both ways (not equal to). Figure 88 represents a sample one way analysis where the null hypothesis states there is no difference and the alternate hypothesis states the sample mean is less than the population mean. If the hypothesis was there was just a difference (non-directional), the rejection area would have been on both ends of the curve.

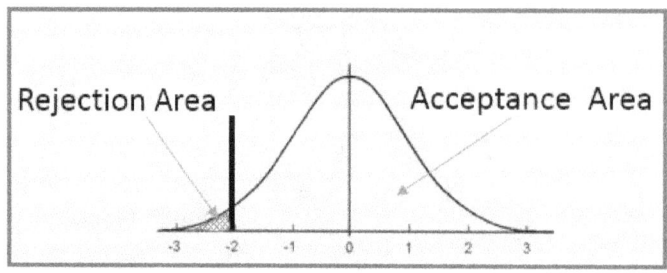

Figure 88: Sample of Rejection Area

The probability of a Type I error should also be stated. These steps are summarized in Figure 89:

Alternate Hypothesis	0.05	0.02	0.01
$\mu - \mu_0 > 0$	1.64	2.06	2.33
$\mu - \mu_0 < 0$	-1.64	-2.06	-2.33
$\mu - \mu_0 \neq 0$	± 1.96	± 2.33	2.58

Remember, μ_0 is the value of the true mean

Figure 89: Probability of Type I Error

The central limit theorem can be used with tests of significance for large samples. A Z-score can be calculated using the following formula as shown in Figure 90:

$$Z = (\overline{X} - \mu_0)/\sigma\sqrt{n}$$

$\overline{X}$ = Point Estimate of μ
μ_0 = Value of True Mean Listed in H_o
σ = Standard Deviation of population
n = Sample Size

Figure 90: Calculation of Z Score

For example, suppose the Director of the local Department of Motor Vehicle has a standard that the average wait time for service is 10 minutes. The manager of the Department collects 40 random wait times from the week's customers to see if the results indicate a higher than normal wait time. In this example, the desired alpha is 0.05. The analysis would be conducted as follows:

Step 1: Collect the data and perform descriptive statistics as shown in previous chapters. The output is shown in Figure 91"

Customer	Wait Time (min)
1	11
2	6
3	10
4	18
5	12
6	8
7	16
8	6
9	12
10	5
11	6
12	13
13	14
14	15
15	17
16	12
17	13
18	12
19	8
20	12
21	14
22	15
23	10
24	7
25	6
26	9
27	9
28	10
29	12
30	17
31	8
32	12
33	16
34	10
35	12
36	15
37	8
38	18
39	7
40	9

Column1	
Mean	11.375
Standard Error	0.679071293
Median	11
Mode	5
Standard Dev	4.294823957
Sample Variance	18.44551282
Kurtosis	-1.236590667
Skewness	0.014852512
Range	13
Minimum	5
Maximum	18
Sum	455
Count	40

Figure 91: Customer Wait Time Descriptive Stats

The sample mean is ~11.38 minutes with a standard deviation of ~4.29 minutes. The standard error of the mean is ~0.68 minutes. Recall that the standard error of the mean is the standard deviation divided by the square root of the number of observations (n). From Table I, an alpha (α) of 0.05 would yield a critical value of Z of 1.64. Note that when using Microsoft® Excel or other software, the obtained critical value is much more precise that what is obtained from a table such as Table 1. See Figure 92 for reminder of how to use Table I.

TABLE I: The Standard Normal Distribution

z	0	0.01	0.02	0.03	0.04	0.05	0.06	0.07	0.08	0.09
0	0	0.00399	0.00798	0.01197	0.01595	0.01994	0.02392	0.0279	0.03188	0.03586
0.1	0.0398	0.0438	0.04776	0.05172	0.05567	0.05966	0.0636	0.06749	0.07142	0.07535
0.2	0.0793	0.08317	0.08706	0.09095	0.09483	0.09871	0.10257	0.10642	0.11026	0.11409
0.3	0.11791	0.12172	0.12552	0.1293	0.13307	0.13683	0.14058	0.14431	0.14803	0.15173
0.4	0.15542	0.1591	0.16276	0.1664	0.17003	0.17364	0.17724	0.18082	0.18439	0.18793
0.5	0.19146	0.19497	0.19847	0.20194	0.2054	0.20884	0.21226	0.21566	0.21904	0.2224
0.6	0.22575	0.22907	0.23237	0.23565	0.23891	0.24215	0.24537	0.24857	0.25175	0.2549
0.7	0.25804	0.26115	0.26424	0.2673	0.27035	0.27337	0.27637	0.27935	0.2823	0.28524
0.8	0.28814	0.29103	0.29389	0.29673	0.29955	0.30234	0.30511	0.30785	0.31057	0.31327
0.9	0.31594	0.31859	0.32121	0.32381	0.32639	0.32894	0.33147	0.33398	0.33646	0.33891
1	0.34134	0.34375	0.34614	0.34849	0.35083	0.35314	0.35543	0.35769	0.35993	0.36214
1.1	0.36433	0.3665	0.36864	0.37076	0.37286	0.37493	0.37698	0.379	0.381	0.38298
1.2	0.38493	0.38686	0.38877	0.39065	0.39251	0.39435	0.39617	0.39796	0.39973	0.40147
1.3	0.4032	0.4049	0.40658	0.40824	0.40988	0.41149	0.41308	0.41466	0.41621	0.41774
1.4	0.41924	0.42073	0.4222	0.42364	0.42507	0.42647	0.42785	0.42922	0.43056	0.43189
1.5	0.43319	0.43448	0.43574	0.43699	0.43822	0.43943	0.44062	0.44179	0.44295	0.44408
1.6	0.4452	0.4463	0.44738	0.44845	0.4495	0.45053	0.45154	0.45254	0.45352	0.45449
1.7	0.45543	0.45637	0.45728	0.45818	0.45907	0.45994	0.4608	0.46164	0.46246	0.46327

Figure 92: Using Table I

The Z score is calculated using the previously mentioned formula.

$$Z = (\overline{X} - \mu_0)/\sigma\sqrt{n}$$

Figure 93: Z Score Formula

In this example, $Z = (11.38 - 10)/ (4.29/6.32) = 2.03$. This is greater than 1.64, so the null hypothesis is rejected and the alternate hypothesis is accepted. See Figure 94. It must be acknowledged that there is a five percent or less chance of committing a type I error. Another way of saying this would be this analysis does not **prove** anything, but it highly suggests that there is a real difference. Likewise, if the calculated Z score was low, it would also not **prove** anything, but would suggest that there is no real difference.

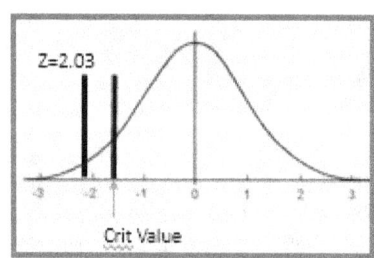

Figure 94: Critical vs. Calculated Value

In the previous example, what would happen if a non-directional alternative hypothesis was proposed? For example, instead of asking if the wait time was longer than 10 minutes, the question may be are the wait times different than 10 minutes? This would indicate a non-directional alternative hypothesis. Using the same alpha of 0.05, the critical value would be 1.96. From Table I, the corresponding area under the curve for 1.96 is 0.475, which means that .025 is to the right and left of 1.96. See Figure 95:

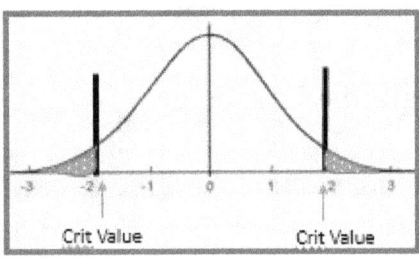

Figure 95: Non-directional Alternative Hypothesis

Problems Chapter 12: (for all tests, assume alpha to be 0.05)

1. There is interest in determining if the number of baseball fans who buy hot dogs is higher on Sundays than on Saturdays. You know the average number of hot dogs sold on Saturday's average 1500 hot dogs. Descriptive statistics are performed on 31 Sunday's and find the average to be 1650 with a standard deviation of 400 hot dogs sold. Is there a statistical difference?

2. There is interest in determining if the number of customer's rate a hotel 5 stars differs by state. You know the average number of 5 star ratings in Louisiana averages 20 hotels/per focus group. Descriptive statistics are performed on 40 focus groups in Virginia. The average number of 5 stars ratings was 28 with a standard deviation of 2 "5 star ratings". Is there a statistical difference?

3. There is interest in determining if the number of football players drafted into the NFL who attended the University of Georgia differs from Georgia Tech. The average per year from Georgia is 4.7 per year. Descriptive statistics are performed on the last 35 years of Georgia Tech and reveals 4.5 are drafted per year with a standard deviation of 0.8 draftees. Is there a statistical difference?

4. Perform the analysis for #3 above with the following difference. You are only interested if Georgia produces more draftees per year.

Solutions Chapter 12:

1. Z = 2.08, Critical value is 1.64, therefore conclusion is that evidence suggest there is a difference in the number of hotdogs sold. The formula to calculate this is "=(1650 - 1500)/(400/31^0.5)"

2. Z = 25.3, and the Critical value is 1.96, therefore conclusion is that evidence suggest there are more 5 star ratings in Virginia than in Louisiana. The formula to calculate this is "=(28-20)/(2/40^0.5)"

3. Z = 1.48, Critical value is 1.96, therefore conclusion is that evidence suggest there are not more draftees from Georgia vs. Georgia Tech. The formula to calculate this is "=(4.7-4.5)/(0.8/35^0.5)"

4. Z=1.48, Critical value is 1.64, so the conclusion is there are no statistical differences between the number of draftees from Georgia vs. Georgia Tech.

CHAPTER 13: TESTING FOR SIGNIFICANT DIFFERENCES – SMALL SAMPLES

When testing for significance in small samples (<30 observations), the option of assuming the central limit theorem is valid do not exist. Recall that the t-distribution can be used for small samples and the t-statistic can be calculated using the formula in Figure 96:

$$t = (\overline{X} - \mu) / S/\sqrt{n}$$

Figure 96: t-Statistic Calculation

If the potential difference is non-directional and the calculated t-statistic exceeds or is less than $\pm t_{\alpha/2,\, n-1t}$, the null hypothesis is rejected. If the potential difference is one-directional and the calculated t-statistic exceeds or is less than $\pm t_{\alpha,\, n-1t}$, the null hypothesis is rejected. In other words, when only examining for differences in one direction, alpha is used. When looking in both directions, alpha divided by two is used.

For example, assume that a basketball coach would like to know if his team scores less than the historical average of the teams in his league. So far this year, the points scored in 12 games played are as follows in Figure 97:

Points Scored
100
120
110
105
98
97
96
100
101
102
100
101

Figure 97: Basketball Points Scored

The historical average for points scored is 105. Therefore, the null hypothesis is – $H_0: \mu = 105$ points and the alternate hypothesis is $H_1: \mu \neq 105$ points. The alpha for this problem is 0.05. The first step to solving this problem is to find the critical value from Table 2. The degrees of freedom (n-1) is 11. Since we are interested in a non-directional analysis, the area of rejection in both sides of the curve is 0.025. See Figure 98:

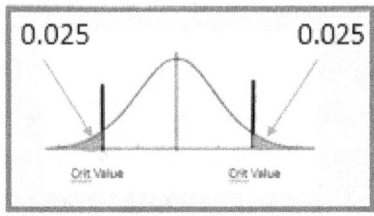

Figure 98: Areas of Rejection of Null Hypothesis

A quick examination of Table 2 will reveal a critical value of 2.2010 for degrees of freedom of 11 and alpha 0.025. See Figure 99:

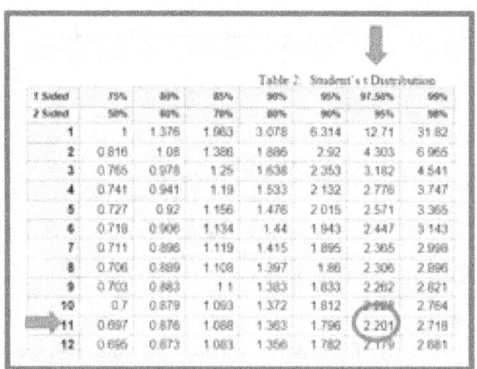

Figure 99: Critical Value from Table 2

The critical value will be lower as the sample size increases. Notice in the current example what occurs if the sample size increased to 13 and the degrees of freedom is 12. Notice that the number is slightly smaller (2.179). This indicates that a smaller calculated t value would be required to suggest a significant difference exists.

The next step is to determine the t-value by calculating descriptive statistics. See Figure 100:

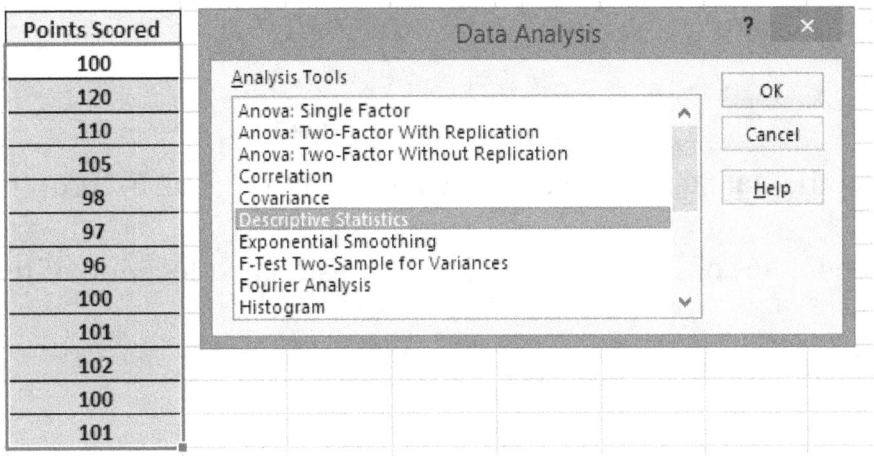

Figure 100: Descriptive Stats

The following results should appear as shown in Figure 101:

Figure 101: Descriptive Stats II

From this data, a mean of 102. 5 is observed. This is certainly below the expected historical mean of 105, but is there any significant difference. The t-value calculation is based on the previously mentioned formula and is shown again in Figure 102:

$$t = (\overline{X} - \mu)/S/\sqrt{n}$$

Figure 102: t-Statistic Calculation

In this case, the statistic can be solved as follows: t = (102.5-105)/ (6.64/Sq. Rt of 12) or -2.5/ (6.64/3.46) or 2.5/1.91 or -1.31. The estimated location of the t value is shown as follows by the "x".

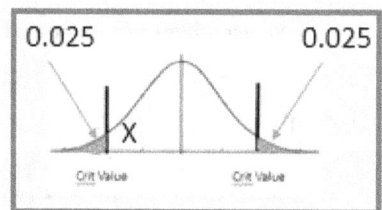

Figure 103: Estimated Location of t-Statistic

The conclusion is that because our calculated t value falls within the nonrejection region, the null hypothesis cannot be rejected. The coach must conclude that statistically speaking, there is no difference between this year's team and last year's team. If the question had been one direction (e.g. did the team score less), a different critical t-value would have been required. The critical value in that case would have been 1.796. Since 1.31 would still fall in the nonelection region, the same conclusion would be drawn and the coach would conclude that his team does not score statistically less than the historical average.

Chapter 13: Problems

1. Using an alpha of 0.05, determine if the following ten samples of cars sold per month differ from the historical average of 8 cars sold per month:

   4,7,8,9,6,7,8,9,6,7

2. Use the same criteria above with the exception you only want to know if the numbers represent a lower number of cars sold.

3. A farmer wants to understand the effect of this year's colder than normal summer on tomatoes per plant in July. The historical average is 10. The farmer counts his eight tomato plants and finds the following: 13,3,10,10,12,5,6,7. Alpha is 0.05 in this example.

4. Solve using the exact same criteria with the exception that Alpha is 0.01.

Chapter 13: Solutions

1. T crit for df=9, alpha is 0.05, non-directional so use alpha/2 or 0.025, so T crit is 2.2622. The calculated t value is (7.1-8)/ (1.13/square root of 10) or -0.9/(1.13/3.16) or 2.52. This result is in the rejection region, we reject the null hypothesis and conclude there is a difference in the number of cars sold.

2. T crit for df=9, alpha is 0.05, directional so use alpha or 0.050, so T crit is 1.183. The calculated t value is (7.1-8)/ (1.13/square root of 10) or -0.9/ (1.13/3.16) or 2.52. This result is in the rejection region, we reject the null hypothesis and conclude there is a difference in the number of cars sold.

3. T crit for df=7, alpha is 0.05, non-directional so use alpha/2 or 0.025, so T crit is 2.365. The calculated t value is (8.25-10)/ (3.53/square root of 8) or -1.75/ (3.53/2.83) or -2.18. This result is in the non-rejection region, we accept the null hypothesis and conclude there is a difference in the number of tomatoes sold.

4. T crit for df=7, alpha is 0.01, non-directional so use alpha/2 or 0.005, so T crit is 3.495. The calculated t value is (8.25-10)/ (3.53/square root of 8) or -1.75/ (3.53/2.83) or -2.18. This result is in the non-rejection region, we accept the

null hypothesis and conclude there is a difference in the number of tomatoes sold.

CHAPTER 14: COMPARING TWO MEANS – TWO SAMPLE Z TEST

There are times when the interest is not in comparing a sample to the larger population, rather, rather, the interest is in the direct comparison of two samples that come from two different populations.

This chapter will cover basic types of tests. Each of these tests, summarized in the flowchart (see Figure 104), can be performed easily with the use of Microsoft® Excel.

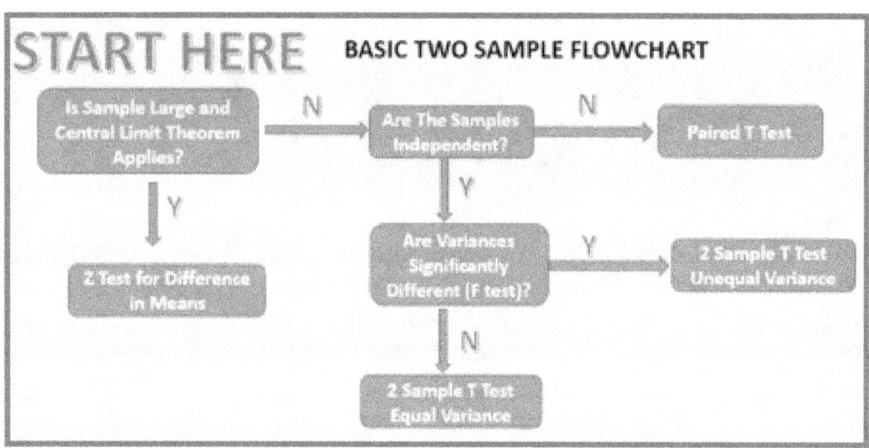

Figure 104: Basic Two Sample Flowchart

In these situations, the development of the hypothesis is the same as in previous chapters. Using the flowchart as a template, the first step in the process of comparing two samples is to determine if the sample is large enough (>30 samples) for the central limit theorem to apply. If the answer is "yes" then the Z test for the difference of means may be an appropriate test. In practice, this test is not used often because the population standard deviations, which are needed to conduct the analysis, are unknown or hard to estimate. The Z test is calculated as shown in Figure 105:

$$z = \frac{\bar{x}_1 - \bar{x}_2}{\sqrt{\frac{\sigma_1^2}{n_1} + \frac{\sigma_2^2}{n_2}}}$$

Figure 105: Z Test Calculation

In this formula, $\bar{x}_1$ and $\bar{x}_2$ are the means of the two samples, n_1 and n_2 are the sample sizes, and σ_1 and σ_2 are the standard deviations of the populations. The following is an example of where the Z test could be used:

Ferritin is an analyte in the blood that measures iron storage. Assume that adult male's ferritin levels have a standard deviation of 50 ng/ml and female ferritin levels have a standard deviation of 35 ng/ml. The researcher takes a random sample of 200 male and 150 females. The resulting mean levels of 280 ng/ml for males and 180 ng/ml for females were obtained. Should the researcher conclude that there is a difference in means between male and female ferritins? The null hypothesis is there is no difference in means and the alternate hypothesis is there is a difference in means. The Z statistic is calculated as follows in Figure 106:

$$\frac{230-180}{\sqrt{\frac{50^2}{200} + \frac{35^2}{150}}} \Rightarrow \frac{50}{\sqrt{\frac{2500}{200} + \frac{1225}{150}}} \Rightarrow \frac{50}{\sqrt{12.5 + 8.17}} \Rightarrow \frac{50}{\sqrt{20.67}} \Rightarrow \frac{50}{4.55} \Rightarrow 10.9$$

Figure 106: Z Statistic Calculation

The result is 10.9. From Table 2, it is apparent there must be a difference as the highest Z score listed is 3.09. The resultant Z score of 10.9 from the above example is certainly greater than 3.09. The conclusion is the data suggests there is a statistically significant difference. Microsoft® Excel is not limited by the confines of a table and will give an exact probability.

Consider the following example of the ages of 31 individuals in two different groups. Are the ages significantly different? Assuming that the central limit theorem applies due to the sample size (31), the use of Microsoft® Excel to compare the means of two groups by a Z test is appropriate. Assume that the known variance is 2.1 for Group 1 and 2.3 for Group 2 and alpha is 0.05. The first step is to enter or import the data into Microsoft® Excel and select "Z-Test Two Sample for Means" from the data analysis function. See Figure 107:

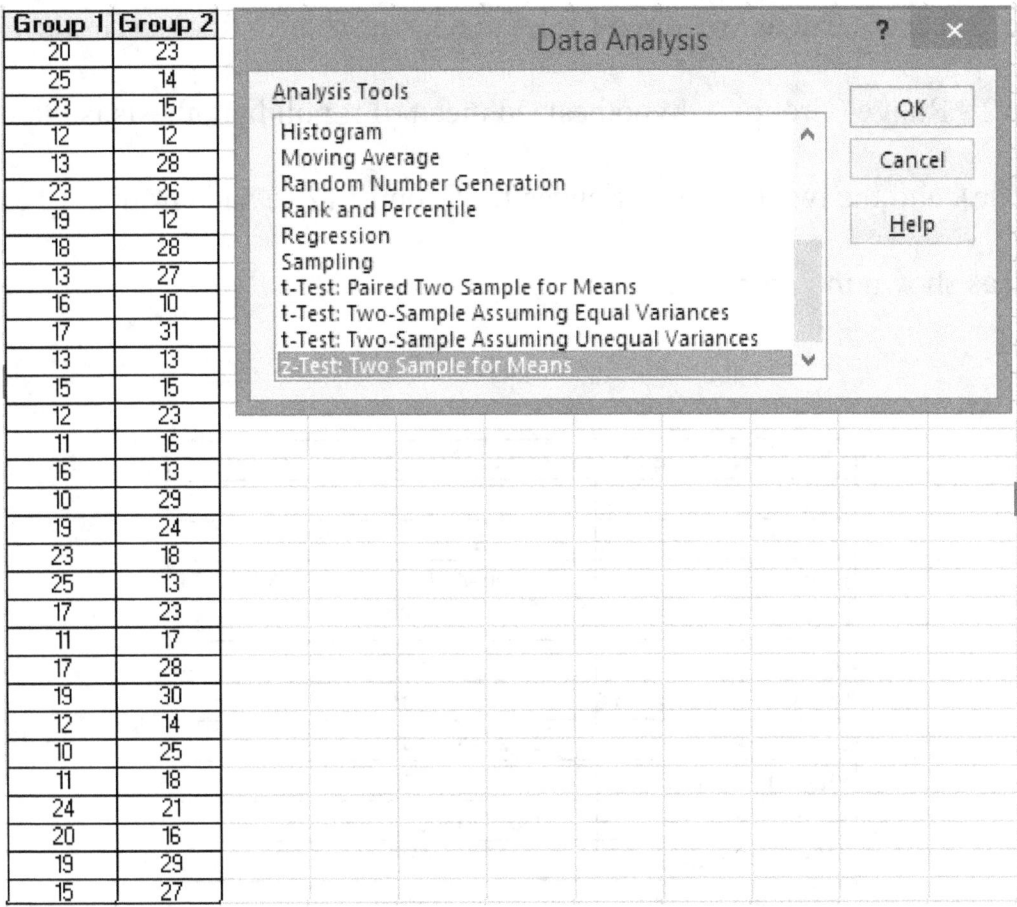

Figure 107: Z-Test - 2 Sample for Means

Next, a dialogue box will appear that will ask for input. See Figure 108. Note that the variance must be known.

Figure 108: Z-Test 2 Sample Input

The five areas that are required for input are Variable 1- Range (Group 1), Variable 2 - Range (Group 2), hypothesized mean (0 if null hypothesis is there is no diffference), and the two known variances for Group 1 (2.1) and Group 2 (2.3). The output is as shown in Figure 109:

z-Test: Two Sample for Means		
	Variable 1	Variable 2
Mean	16.70967742	20.58064516
Known Variance	2.1	2.3
Observations	31	31
Hypothesized Mean D	0	
z	-10.27481618	
P(Z<=z) one-tail	0	
z Critical one-tail	1.644853627	
P(Z<=z) two-tail	0	
z Critical two-tail	1.959963985	

Figure 109: Z-Test 2 Sample Output

The calculated Z score is -10.27 which is far less than the critical values for either the one or two tailed test. The researcher should conclude the evidence suggests there is a signficant difference in means.

Without showing all the steps as in the previous example, consider the following different example of ages of 31 individuals in two different groups. Are the ages significantly different? Assuming that the central limit theorem applies due to the sample size (31), the use of Microsoft® Excel to compare the means of two groups by a Z test is appropriate. Assume that the known variance is 5.1 for Group 1 and 5.3

for Group 2 and alpha is 0.05. The first step is to enter or import the data into Microsoft® Excel and select "Z-Test Two Sample for Means" from the data analysis function. See Figure 110:

Group 1	Group 2
18	11
18	11
18	13
14	12
18	11
16	19
15	16
11	16
14	16
11	12
14	11
17	11
14	19
13	12
11	18
11	12
12	11
13	13
10	12
10	18
18	14
16	13
11	15
20	17
18	12
14	13
12	11
12	11
18	16
13	12
15	17

Figure 110: Ages of Two Groups

Analysis using the Z-test two sample means yeilds the following. See Figure 111:

z-Test: Two Sample for Means		
	Variable 1	Variable 2
Mean	14.3548387	13.70967742
Known Variance	5.1	5.3
Observations	31	31
Hypothesized Mean	0	
z	1.1138648	
P(Z<=z) one-tail	0.1326686	
z Critical one-tail	1.64485363	
P(Z<=z) two-tail	0.26533719	
z Critical two-tail	1.95996398	

Figure 111: Z-Test 2 Sample Output

The overall Z sscore is 1.11 which does not exceed the critical Z score for either the one or two tailed tests and indicates there is no evidence to support a statistical difference in means. A review of the data shows that not only are the means closer together but there is also much greater variance than in the previous example. What would happen if the exact same data were used but the known variance was 0.1 for both groups? The output is shown in Figure 112:

z-Test: Two Sample for Means		
	Variable 1	Variable 2
Mean	14.35483871	13.70967742
Known Variance	0.1	0.1
Observations	31	31
Hypothesized Mean Difference	0	
z	8.032193289	
P(Z<=z) one-tail	4.44089E-16	
z Critical one-tail	1.644853627	
P(Z<=z) two-tail	8.88178E-16	
z Critical two-tail	1.959963985	

Figure 112: Z-Test 2 Sample Output

Notice that given the exact same data set, the simple change of the known variance completely changes the answer (Z = 8.03). The less variation that exists in the data set, the easier it is to demonstrate there is a probable difference.

Chapter 14 Questions: The following data should be used in solving for differences in the following problems.

Group A	Group B	Group C	Group D	Group E
8	16	11	39	30
6	12	7	10	25
20	25	39	27	31
40	16	40	38	39
36	13	40	24	37
23	41	34	6	9
5	26	9	20	32
21	11	4	52	37
7	17	39	26	17
39	11	29	41	18
40	3	12	10	25
10	28	5	30	40
36	25	20	2	5
28	3	34	2	40
25	1	12	18	33
34	15	0	49	4
12	38	30	47	16
22	32	22	8	14
1	24	37	46	15
22	16	6	40	8
22	25	28	32	14
39	39	8	24	31
33	38	21	38	36
35	31	47	22	25
29	42	39	30	19
5	8	23	12	1
26	41	45	38	40
4	8	7	4	6
11	29	34	19	12
4	24	42	42	17
31	10	4	5	36

Assume the central limit theorum applies and alpha is 0.05 for each Group. The known variations are 0.1 for Group A, 0.2 for Group B, 0.3 for Group C, 0.4 for Group D, and 15 for Group E. Solve for two tail differences.

1. Determine if there are difference in means of Group A vs. Group B.

2. Determine if there are difference in means of Group A vs. Group C.

3. Determine if there are difference in means of Group A vs. Group D.

4. Determine if there are difference in means of Group A vs. Group E.

5. Determine if there are difference in means of Group B vs. Group C.

6. Determine if there are difference in means of Group B vs. Group D.

7. Determine if there are difference in means of Group B vs. Group E.

8. Determine if there are difference in means of Group C vs. Group D.

9. Determine if there are difference in means of Group C vs. Group E.

10. Determine if there are difference in means of Group D vs. Group E.

Chapter 14: Solutions.

1. $Z = 1.97$, therefore significant difference.

2. $Z = -15.335$, therefore significant difference.

3. $Z = -32.26$, therefore significant difference.

4. $Z = 1.75$, therefore no significant difference.

5. $Z = -15.24$, therefore significant difference.

6. $Z = -30.83$, therefore significant difference.

7. $Z = -2.02$, therefore significant difference.

8. $Z = -15.67$, therefore significant difference.

9. $Z = 0.73$, therefore no significant difference.

10. $Z = 4.07$, therefore significant difference.

CHAPTER 15: COMPARING TWO MEANS – T TEST FOR RELATED SAMPLES (PAIRED T TEST)

From the flowchart in Chapter 14, if the sample size is less than 30 and the central limit theorem does not apply, the next step is to determine if the samples are independent. If the samples are not independent, then the t test for related samples or paired t test is appropriate. In this chapter, the observations are either matched or are repeated observations obtained from the same item of interest. We assume that there is an .underlying similarity of variables that may affect the observations. Any differences identified should be due to other unknown variables. The analysis is very similar to one sample t-tests (Chapter 13).

To understand how Microsoft® Excel is actually performing the analysis, it is instructive to understand a simple example of how this test is performed manually. The following example uses Microsoft® Excel perform simple mathematical calculations, but not using Microsoft® Excel to perform the analysis. Consider the following example of SAT math scores taken by the same students for two different exam dates. The first step is to list the test scores side by side and calculate the difference between the two "matched" scores and also square the differences. This is shown in Figure 113:

TEST 1	TEST 2	Diff	Diff Squared
730	690	40	1600
550	580	-30	900
600	640	-40	1600
590	590	0	0
600	580	20	400
700	650	50	2500
680	630	50	2500
500	470	30	900
540	550	-10	100
780	750	30	900

Figure 113: Comparison of Test Scores

The next step is to sum the differences and sum the differences squared. See Figure 114:

TEST 1	TEST 2	Diff	Diff Squared
730	690	40	1600
550	580	-30	900
600	640	-40	1600
590	590	0	0
600	580	20	400
700	650	50	2500
680	630	50	2500
500	470	30	900
540	550	-10	100
780	750	30	900
	SUM	140	11400

Figure 114: Comparison of Test Scores II

The t statistic is calculated as follows where N is the number of matched pairs and d is the individual differences between matched pairs. See Figure 115:

$$t = \frac{\frac{\Sigma d}{N}}{\sqrt{\frac{\Sigma d^2 - \frac{(\Sigma d)^2}{N}}{N(N-1)}}}$$

Figure 115: t-Statistic Calculation

Specifically for this example, the t statistic is calculated as follows in Figure 116:

$$t = \frac{\frac{140}{10}}{\sqrt{\frac{11400 - \frac{(140)^2}{10}}{(10)(9)}}} = t = \frac{14}{\sqrt{\frac{11400 - 1960}{90}}} = t = \frac{14}{10.24} = 1.37$$

Figure 116: t-Statistic Calculation II

From Table 2, it is determined that t critical value for alpha of 0.05 and degrees of freedom of n-1(9) is 2.262. The calculated t statistic of 1.37 is in the area of nonrejection of the null hypothesis, so the conclusion is there is no statistical difference in SAT math scores.

Microsoft® Excel solves paired t testing problems quickly. After the paired data has been entered into a worksheet, the "t-test: Paired Two Sample for Means" should be selected from the data analysis tab. See Figure 117:

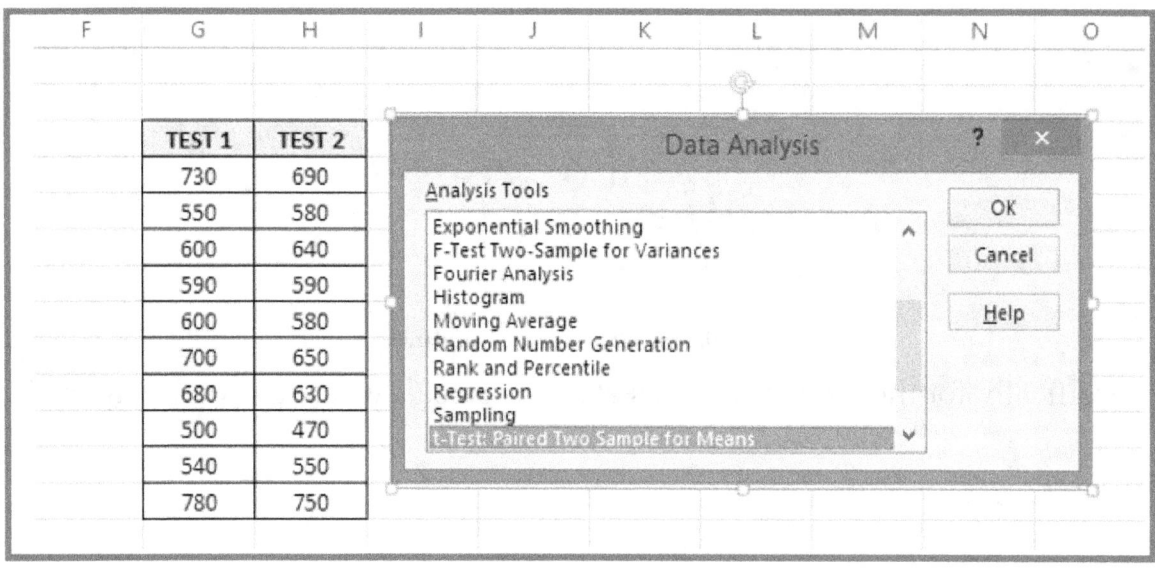

Figure 117: Paired t Test

Next, the two variable ranges are entered. See Figure 118:

Figure 118: Paired t Test II

The result of the analysis are shown in Figure 119. Note that t statistic and the t critical are very close to the result obtained from calculation by hand.

t-Test: Paired Two Sample for Means		
	Variable 1	Variable 2
Mean	627	613
Variance	8290	6023.333333
Observations	10	10
Pearson Correlation	0.93856284	
Hypothesized Mean	0	
df	9	
t Stat	1.366983565	
P(T<=t) one-tail	0.102399744	
t Critical one-tail	1.833112933	
P(T<=t) two-tail	0.204799489	
t Critical two-tail	2.262157163	

Figure 119: Paired t Test Output

Chapter 15: Problems

1. Determine if there are any differences in the means of the following paired samples. Assume alpha is 0.05 and the interested difference is bi-directional (two tailed test).

Group 1	Group 2
1	0
2	3
3	2
4	5
5	3
6	7
7	6
8	9
9	9
10	10

2. Determine if there are any differences in the means of the following paired samples. Assume alpha is 0.05 and the interested difference is bi-directional (two tailed test).

Group 1	Group 2
17	18
21	22
25	24
29	28
33	29
37	33
32	36
28	24
22	18
18	22

3. Determine if there are any differences in the means of the following paired samples. Assume alpha is 0.05 and the interested difference is one-directional (one tailed test).

Group 1	Group 2
100	92
150	165
200	146
150	149
100	67
50	61
100	152
150	141
100	111
50	49

4. Determine if there are any differences in the means of the following paired samples. Assume alpha is 0.01 and the interested difference is one-directional (one tailed test).

Group 1	Group 2
5	6
5	6
5	6
5	6
5	6
4	6
5	5
5	6
4	6
5	6

Chapter 15: Solutions

1. No statistical difference is noted (t stat is 0.29 and t crit is 2.262).

t-Test: Paired Two Sample for Means

	Variable 1	Variable 2
Mean	5.5	5.4
Variance	9.166666667	11.37777778
Observations	10	10
Pearson Correlation	0.946547254	
Hypothesized Mean Difference	0	
df	9	
t Stat	0.287347886	
P(T<=t) one-tail	0.390176168	
t Critical one-tail	1.833112933	
P(T<=t) two-tail	0.780352336	
t Critical two-tail	2.262157163	

2. No statistical difference is noted (t stat is 0.78 and t crit is 2.262).

t-Test: Paired Two Sample for Means

	Variable 1	Variable 2
Mean	26.2	25.4
Variance	45.06666667	36.26666667
Observations	10	10
Pearson Correlation	0.877281224	
Hypothesized Mean Difference	0	
df	9	
t Stat	0.784464541	
P(T<=t) one-tail	0.226459873	
t Critical one-tail	1.833112933	
P(T<=t) two-tail	0.452919746	
t Critical two-tail	2.262157163	

3. No statistical difference is noted (t stat is 0.19 and t crit is 1.83).

t-Test: Paired Two Sample for Means

	Variable 1	Variable 2
Mean	115	113.3
Variance	2250	1861.566667
Observations	10	10
Pearson Correlation	0.806491122	
Hypothesized Mean Difference	0	
df	9	
t Stat	0.188835794	
P(T<=t) one-tail	0.427205563	
t Critical one-tail	1.833112933	
P(T<=t) two-tail	0.854411126	
t Critical two-tail	2.262157163	

4. A statistical difference is noted (t stat is -6.13 and t crit is 2.82).

t-Test: Paired Two Sample for Means

	Variable 1	Variable 2
Mean	4.8	5.9
Variance	0.177777778	0.1
Observations	10	10
Pearson Correlation	-0.166666667	
Hypothesized Mean Difference	0	
df	9	
t Stat	-6.12794616	
P(T<=t) one-tail	8.66386E-05	
t Critical one-tail	2.821437925	
P(T<=t) two-tail	0.000173277	
t Critical two-tail	3.249835542	

CHAPTER 16: TESTING FOR SIGNIFICANT DIFFERENCES – INDEPENDENT SMALL SAMPLES

If the number of sample observations is small and the samples are not related (independent), the two most common testing alternatives are the t-tests of equal and unequal variances. Before examining these two tests, the method for determining if variances are or are not equal must be addressed.

The method to determine if the variances, drawn from samples of two independent populations, are different is based on a ratio. This ratio is calculated from the variances of the two samples. It is important to understand the underlying assumption in this methodology is the assumption that the independent population from which both samples are drawn is normally distributed. The ratio of the variance of sample 1 divided by sample 2 is called the *F test statistic*. F statistic has a characteristic distribution with associated degrees of freedom of each sample minus one (n_1-1, n_2-1). Please refer to other statistical textbooks for a complete list of *F* statistic tables. This chapter will not deal with manual calculations as Microsoft® Excel will perform this function.

Similar to previously discussed hypothesis testing, the null hypothesis for comparing two variances is that there is no difference in variance. The alternate hypothesis is that there is a difference in variance. The calculated ratio of the variance of sample one and sample two is compared with the F critical value with the appropriate degrees of freedom to determine if the variances are indeed different or equal.

Using Microsoft® Excel, the potential difference of variance in two datasets can easily be compared. The first step is to input the data into Microsoft® Excel. Consider the following two groups shown in Figure 120. Assuming alpha is 0.05, determine if there is a difference in variance in the two groups.

Group 1	Group 2
10	8
10	8
10	8
10	8
10	7
9	7
8	7
10	7
9	8
9	8

Figure 120: Data for F-Test Comparison

Next, select the "F-Test Two-Sample for Variances" function from the data analysis tab. See Figure 121:

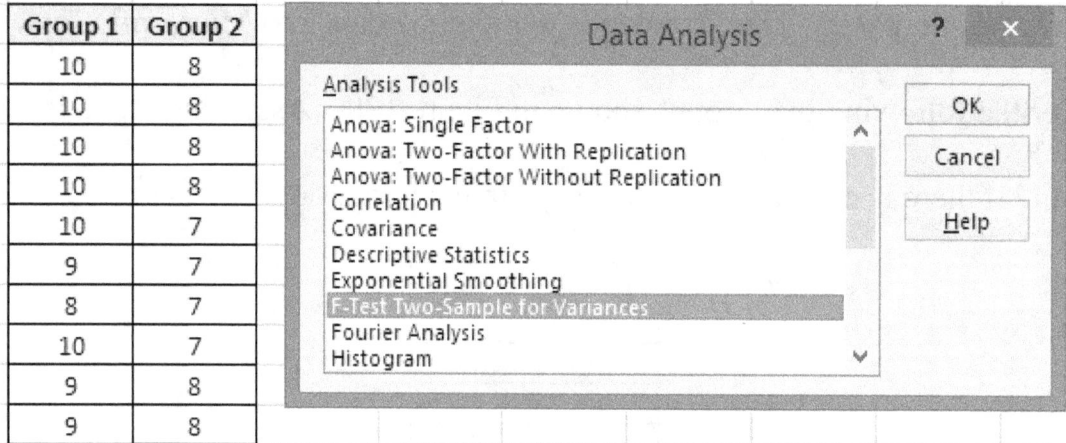

Figure 121: F Test Setup

After inputting the two variable ranges, Microsoft® Excel produces the following analysis. See Figure 122.

F-Test Two-Sample for Variances

	Variable 1	Variable 2
Mean	9.5	7.6
Variance	0.5	0.266666667
Observations	10	10
df	9	9
F	1.875	
P(F<=f) one-tail	0.181421	
F Critical one-tail	3.178893	

Figure 122: F Test Output

Note that the calculated F statistic is less than the F critical and therefore the conclusion is that the variances are equal. It is also noteworthy to point out that the F-test was not required when performing an analysis on two paired or highly related samples. This is due to the inherent similarity of the samples. In other words, there is no need to compare the variances of the two samples when their similarity, by nature, indicates their variances should be similar.

The final two tests that will be discussed in this chapter are the separate variance t-tests, for when the variances are not equal and the pooled variance t-test, for when the variances are equal. The calculation for the t statistic looks intimidating, but in fact, it is a simple calculation. The formula for the t statistic calculation is as follows along with the calculation for pooled variance (S^2_p). See Figure 123:

$$t = \frac{(\bar{X}_1 - \bar{X}_2)}{\sqrt{S_p^2 \left(\frac{1}{N_1} + \frac{1}{N_2}\right)}}$$

$$S_p^2 = \frac{(N_1 - 1) s_1^2 + (N_2 - 1) s_2^2}{N_1 + N_2 - 2}$$

t statistic calculation for 2 sample pooled variance (variances equal)

Figure 123: t Statistic and S^2_p Calculation

Consider the following example. Assume that descriptive statistics have been calculated on two independent samples. The *F* test indicates that the variances are equal. For this problem, alpha is 0.05. The pertinent information to solve for t is shown in Figure 124:

	Group 1	Group 2
n	20	20
mean	50	50
variance	350	200

Figure 124: Sample Data

The first step is to calculate the pooled variance. See Figure 125:

$$S^2p = \frac{19(350) + 19(200)}{20 + 20 - 2} = \frac{6650 + 3800}{38} = 275$$

Figure 125: Pooled Variance Calculation

The t statistic is calculated as shown in Figure 126:

$$t = \frac{(50 - 70)}{\sqrt{275 \left(\frac{1}{20} + \frac{1}{20}\right)}}$$

$$t = \frac{-20}{\sqrt{27.5}}$$

$$t = \frac{-20}{5.244} = -3.81$$

Figure 126: t Statistic Calculation

From Table 2, it can be seen that tcrit for t_{40} is 2.021. The degrees of freedom for this example is 38 which is not in the table. The next smallest degrees of freedom in Table 2 is 30, so the conservative approach is to use 30 as the degrees of freedom, which is 2.042. The calculated t value is -3.81, therefore the null hypothesis is rejected and the conclusion is: there is a probable statistical difference in the means of the two samples.

Microsoft® Excel easily perform the pooled variance t-test. Consider the following example of 15 independent samples. The variance is considered equal. Alpha is 0.05. See Figure 127:

Group 1	Group 2
20	25
21	25
22	24
21	25
20	23
19	24
21	25
23	26
21	24
19	23
20	22
20	23
20	24
21	25
21	25

Figure 127: Pooled Variance Data

Next, the "t-Test: Two-Sample Assuming Equal Variances" is selected from the data analysis tab. See Figure 128:

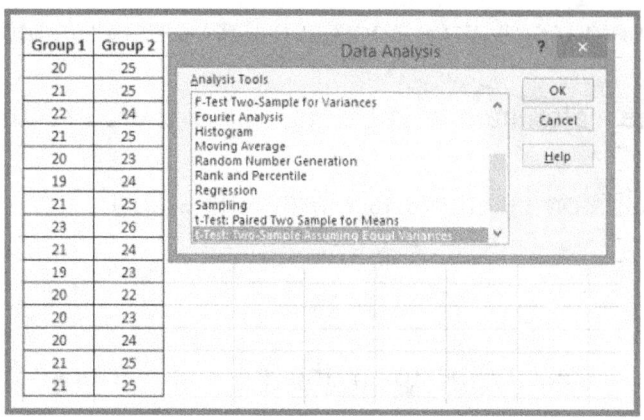

Figure 128: t-Test: Two Sample Assuming Equal Variances Setup

After entering the two variable ranges, the resulting output is as follows in Figure 129:

t-Test: Two-Sample Assuming Equal Variances		
	Variable 1	Variable 2
Mean	20.6	24.2
Variance	1.114286	1.171428571
Observations	15	15
Pooled Variance	1.142857	
Hypothesized Mean Differ	0	
df	28	
t Stat	-9.22226	
P(T<=t) one-tail	2.79E-10	
t Critical one-tail	1.701131	
P(T<=t) two-tail	5.57E-10	
t Critical two-tail	2.048407	

Figure 129: t-Test: Two Sample Assuming Equal Variances Output

The t-stat is much less (-9.22) compared to either one or two tail crit values. The conclusion is there is probable evidence that a statistical difference exists between the means of these two independent samples.

The procedure for determining if there is a significant difference between two samples with separate variances is somewhat different. In this analysis, two separate variances are needed for the computation. The manual computation of the separate variances will not be reviewed in this text as it is complex. The method for performing the t-test for separate variances using Microsoft® Excel will be covered. Consider the example of the following independent samples with unequal variances. Please see Figure 130:

Group 1	Group 2
20	5
21	35
22	20
21	17
20	10
19	21
21	9
23	32
21	15
19	24
20	26
20	3
20	28
21	7
21	28

Figure 130: Independent Samples with Unequal Variances

From the data analysis tab, select the "t-Test: Assuming Unequal Variances". See Figure 131:

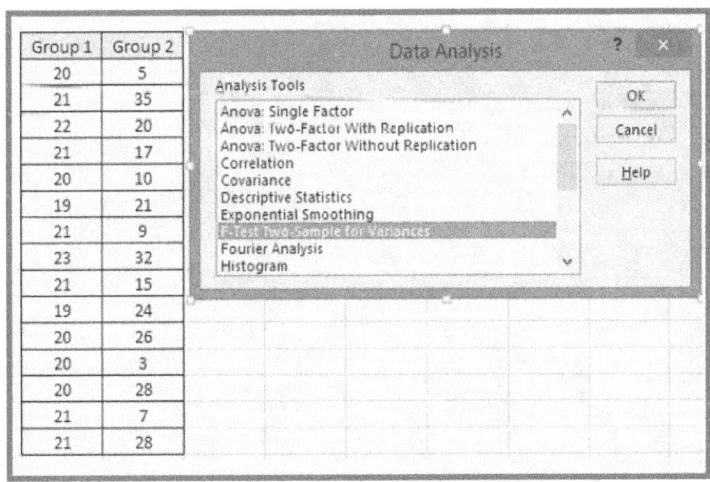

Figure 131: Independent Samples with Unequal Variances Setup

After entering the data for the two variables, the output is as shown in Figure 132:

t-Test: Two-Sample Assuming Unequal Variances		
	Variable 1	Variable 2
Mean	20.6	18.66666667
Variance	1.114285714	104.3809524
Observations	15	15
Hypothesized Mean Diff	0	
df	14	
t Stat	0.72901412	
P(T<=t) one-tail	0.239006188	
t Critical one-tail	1.761310136	
P(T<=t) two-tail	0.478012377	
t Critical two-tail	2.144786688	

Figure 132: Independent Samples with Unequal Variances Output

The calculated value falls within the non-rejection area (t stat = 0.73), so the null hypothesis is accepted and the conclusion is there is no probable significant difference in means of the two samples.

Chapter 16 - Questions

1. Determine if there is a difference in variance in the following two samples. Assume alpha is 0.05.

Group 1	Group 2
7	7
9	5
6	7
2	8
10	2
6	6
1	1
2	8
7	5
10	9

2. Determine if there is a difference in variance in the following two samples. Assume alpha is 0.05.

Group 1	Group 2
7	6
2	6
8	5
6	4
4	5
3	3
7	7
3	7
8	3
7	5

3. Determine if there is a difference in variance in the following two samples. Assume alpha is 0.05.

Group 1	Group 2
9	4
10	5
8	4
8	4
7	5
8	5
7	5
8	4
3	5
1	4

4. Assuming alpha is 0.05 and the variances are equal, determine if there is a significant difference between these two independent samples.

Group 1	Group 2
5	8
5	8
5	8
5	8
5	8
6	8
6	9
6	9
6	9
6	9

5. Assuming alpha is 0.05 and the variances are equal, determine if there is a significant difference between these two independent samples.

Group 1	Group 2
4	3
4	3
4	3
4	3
4	4
4	4
4	4
3	5
4	4
4	3

6. Assuming alpha is 0.05 and the variances are unequal, determine if there is a significant difference between these two independent samples.

Group 1	Group 2
10	7
5	7
10	7
5	7
9	7
4	7
9	7
4	7
9	8
4	7

7. Assuming alpha is 0.05 and the variances are unequal, determine if there is a significant difference between these two independent samples.

Group 1	Group 2
10	1
5	10
10	4
5	4
9	3
4	7
9	8
4	2
9	9
4	1

Chapter 16 – Solutions

1. There is no statistical differences in variances of samples. The calculated F is 1.62.

F-Test Two-Sample for Variances

	Variable 1	Variable 2
Mean	6	5.8
Variance	11.11111111	6.844444444
Observations	10	10
df	9	9
F	1.623376623	
P(F<=f) one-tail	0.24084735	
F Critical one-tail	3.178893104	

2. There is no statistical differences in variances of samples. The calculated F is 2.46.

F-Test Two-Sample for Variances

	Variable 1	Variable 2
Mean	5.5	5.1
Variance	5.166666667	2.1
Observations	10	10
df	9	9
F	2.46031746	
P(F<=f) one-tail	0.098021596	
F Critical one-tail	3.178893104	

3. There is no statistical differences in variances of samples. The calculated F is 27.56.

F-Test Two-Sample for Variances

	Variable 1	Variable 2
Mean	6.9	4.5
Variance	7.655555556	0.277777778
Observations	10	10
df	9	9
F	27.56	
P(F<=f) one-tail	1.68261E-05	
F Critical one-tail	3.178893104	

4. The analysis suggests that a statistically significant difference in means exists (t stat = -12.4).

t-Test: Two-Sample Assuming Equal Variances

	Variable 1	Variable 2
Mean	5.5	8.4
Variance	0.277777778	0.266666667
Observations	10	10
Pooled Variance	0.272222222	
Hypothesized Mean Difference	0	
df	18	
t Stat	-12.4285714	
P(T<=t) one-tail	1.43128E-10	
t Critical one-tail	1.734063607	
P(T<=t) two-tail	2.86257E-10	
t Critical two-tail	2.10092204	

5. The analysis suggests that a statistically significant difference in means does not exist (t stat = 1.23).

t-Test: Two-Sample Assuming Equal Variances

	Variable 1	Variable 2
Mean	3.9	3.6
Variance	0.1	0.488888889
Observations	10	10
Pooled Variance	0.294444444	
Hypothesized Mean Difference	0	
df	18	
t Stat	1.236245076	
P(T<=t) one-tail	0.11612853	
t Critical one-tail	1.734063607	
P(T<=t) two-tail	0.23225706	
t Critical two-tail	2.10092204	

6. The analysis suggests that a statistically significant difference in means does not exist (t stat = 0.23).

t-Test: Two-Sample Assuming Unequal Variances

	Variable 1	Variable 2
Mean	6.9	7.1
Variance	7.211111111	0.1
Observations	10	10
Hypothesized Mean Difference	0	
df	9	
t Stat	-0.233904353	
P(T<=t) one-tail	0.41014657	
t Critical one-tail	1.833112933	
P(T<=t) two-tail	0.820293139	
t Critical two-tail	2.262157163	

7. The analysis suggests that a statistically significant difference in means does not exist (t stat = 1.47).

t-Test: Two-Sample Assuming Unequal Variances

	Variable 1	Variable 2
Mean	6.9	4.9
Variance	7.211111111	11.21111111
Observations	10	10
Hypothesized Mean Difference	0	
df	17	
t Stat	1.473530017	
P(T<=t) one-tail	0.079441859	
t Critical one-tail	1.739606726	
P(T<=t) two-tail	0.158883719	
t Critical two-tail	2.109815578	

CHAPTER 17: TESTING FOR SIGNIFICANT DIFFERENCES – MORE THAN TWO GROUPS – ONE WAY ANALYSIS OF VARIANCE

In previous chapters, differences in two groups have been examined. Many times, there are more than two groups that need to be analyzed. Luckily, many of the same principles discussed in previous chapters can assist in determining differences in the means of more than one group. The first test to be covered is the One Way Analysis of Variance or One Way ANOVA.

The overarching concept to understanding ANOVA is the different ways variance could impact our ability to discern differences in group means. Consider the Figure 133 that shows satisfaction ratings of three different apples. Each apple had five different ratings. What are the different ways variation could exist? Notice that each of the fifteen individual results differ from the overall mean of 7.36. The variation can exist from **within** in the groups (columns) and from **between** the groups (rows).

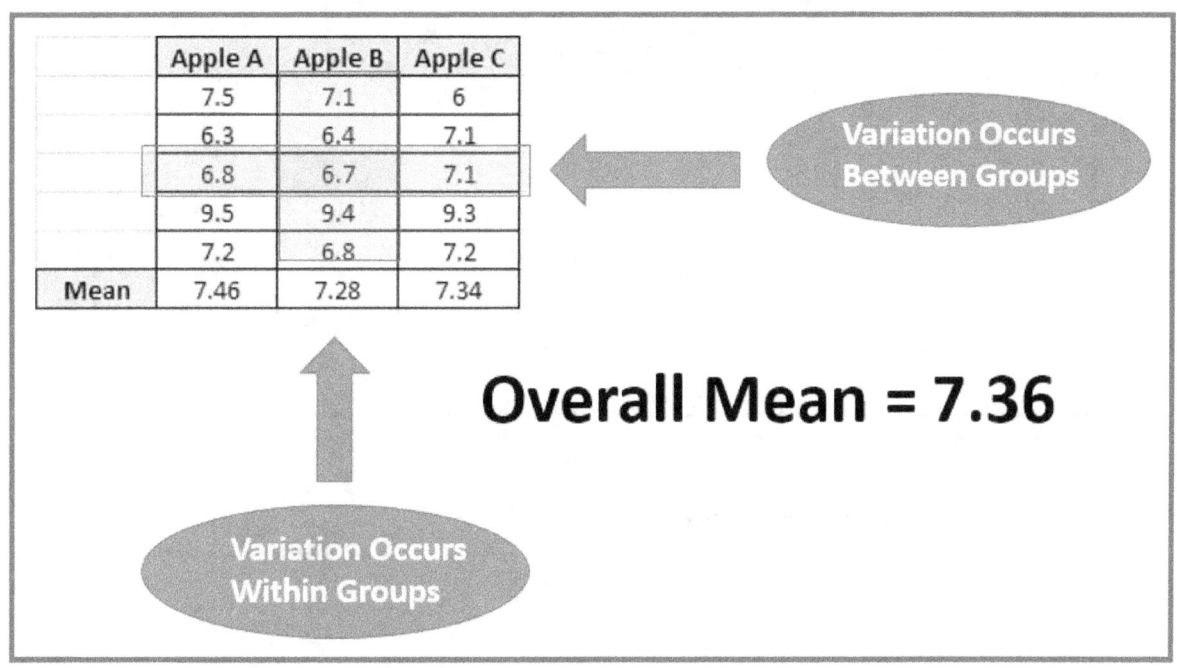

Figure 133: Variation Sources

There are times where variation exists and no significant difference exists between the samples. Consider the following examples of three normally distributed curves of data and notice the overall difference between the three curves. See Figures 134 and 135. In the first figure, there is obvious overlap among all three curves. In the second figure, curve "C" is obviously different than curve "A" and perhaps different than curve "B". The next question is why and how? The ANOVA helps with these questions.

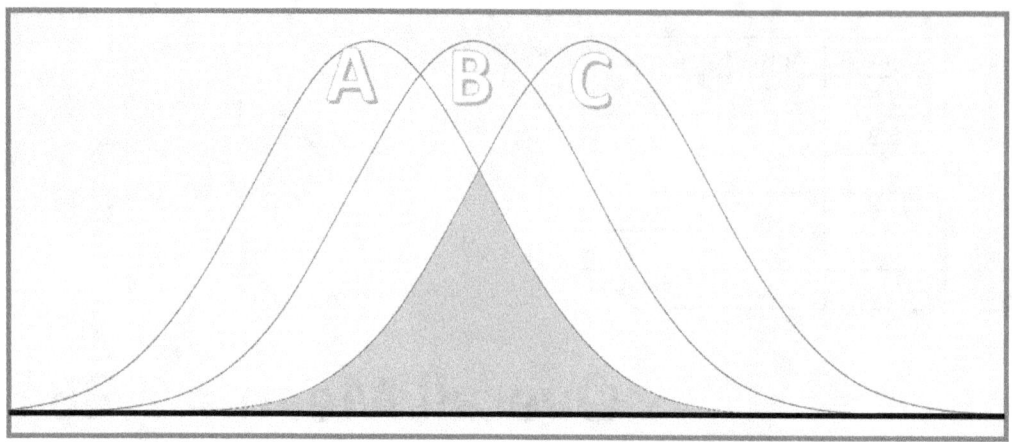

Figure 134: Difference Between Distributions I

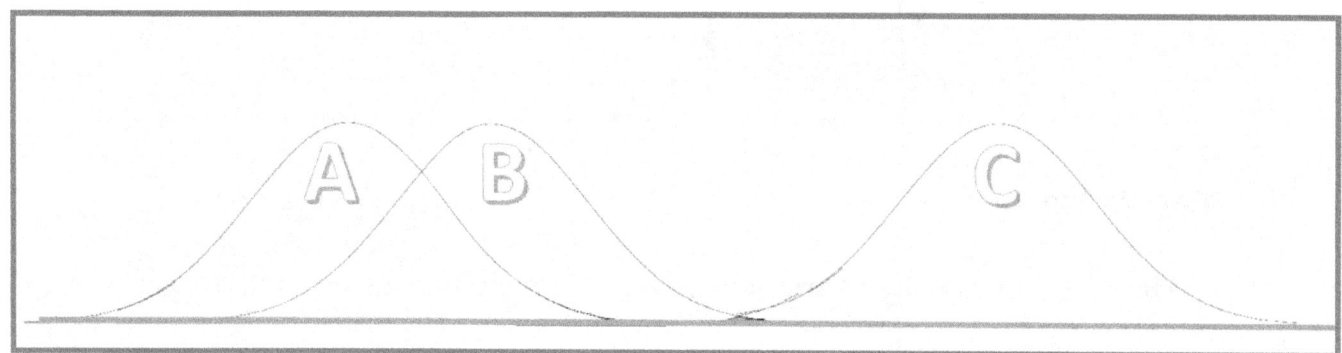

Figure 135: Difference Between Distributions II

After understanding that variation can come from between and within the groups, the next questions is why does variation exist? It is already noted that in the previous example, the means were different for each group. The differences could be because there is truly a difference between the apples. It could also be due to inherent variation (individual differences) within the group of one or more apples. It could also be due to an error on the part of the individual performing the measurement or the inherent error in the measurement tool (experimental error). The variability between the groups can also be caused by a combination of individual differences and experimental error. However, there can also be a unique reason for the variation that is unique for between the groups known as "the treatment effect".

The total overall variability is the sum of variability within each group and between each group. The ANOVA quantifies each of the components of variability and it contributes to the total variability. After determining if there are significant differences in the means of three or more groups, additional methods can be used to determine which specific means differ from other specific means (e.g. mean A is different from mean C, but not different from mean B).

The primary component of the measure of variability is the sum of squares. Recall that the sum of squares was introduced in Chapter 5 when discussing variance. The total variation for all the groups is total sum of squares for all the groups. It is abbreviated SS_{tot} and has an associated degrees of freedom of n-1. The within group variation is total sum of squares within the groups is abbreviated SS_w and has an associated degrees of freedom of n-c where "c" represents the number of groups. The total variation between the groups is total sum of squares between the groups and is abbreviated SS_b and has an associated degrees of freedom of c-1. The last major component of the ANOVA is the mean of squares (MS). The formulas for calculating the mean of squares is as follows in Figure 136:

MEAN SQUARES DEFINITIONS

$$MS_{total} = \frac{SS_{total}}{n-1}$$

$$MS_{Within} = \frac{SS_{within}}{n-c}$$

$$MS_{between} = \frac{SS_{between}}{c-1}$$

Figure 136: Mean of Squares

The null hypothesis for ANOVA is $H_0: \mu_1 = \mu_2 = \mu_3 \ldots = \mu_c$. The alternate hypothesis is there is some difference between the groups. The end goal of the one way ANOVA is a calculated F statistic. The F distribution was introduced in the last chapter. The F statistic follows the F distribution and is calculated as follows in Figure 137:

$$F = \frac{MS_{between}}{MS_{within}}$$

Figure 137: F Statistic Calculation

The closer the F statistic is to one or less, the more likely the null hypothesis will be accepted. This is intuitive because the larger the F statistic, the larger the proportion of the MS_b to the MS_w. In other words, more of the potential difference in means is due to the variance between the groups and not within the group.

For many, the ANOVA is a challenging concept. A different way of looking at the ANOVA, before seeing how the answer is obtained, is to see a sample output from Microsoft® Excel. An example of an ANOVA summary table is as follows in Figure 138:

SUMMARY						
Groups	Count	Sum	Average	Variance		
Column 1	5	37.3	7.46	1.503		
Column 2	5	36.4	7.28	1.467		
Column 3	5	36.7	7.34	1.443		
ANOVA						
Source of Variation	SS	df	MS	F	P-value	F crit
Between Groups	0.084	2	0.042	0.028552	0.971918	3.885294
Within Groups	17.652	12	1.471			
Total	17.736	14				

Figure 138: Example of ANOVA Summary Table

Note that the calculated F statistic (0.028) is far less than the F critical value. This indicates that there is probable evidence to support no difference between any of the means. This ANOVA table came from the example of the three apples.

Now that the end result of an ANOVA has been observed, Microsoft® Excel will be used to manually demonstrate where all these numbers came from. The first part of the ANOVA table is the "SUMMARY" which consists of the count, sum, average, and variance. Each of these four descriptors can be generated with Microsoft® Excel using the descriptive statistics function. For the ANOVA portion, the starting point will be the data. Using the same apple dataset, the ANOVA table can be completed by following these steps. See Figures 139-142:

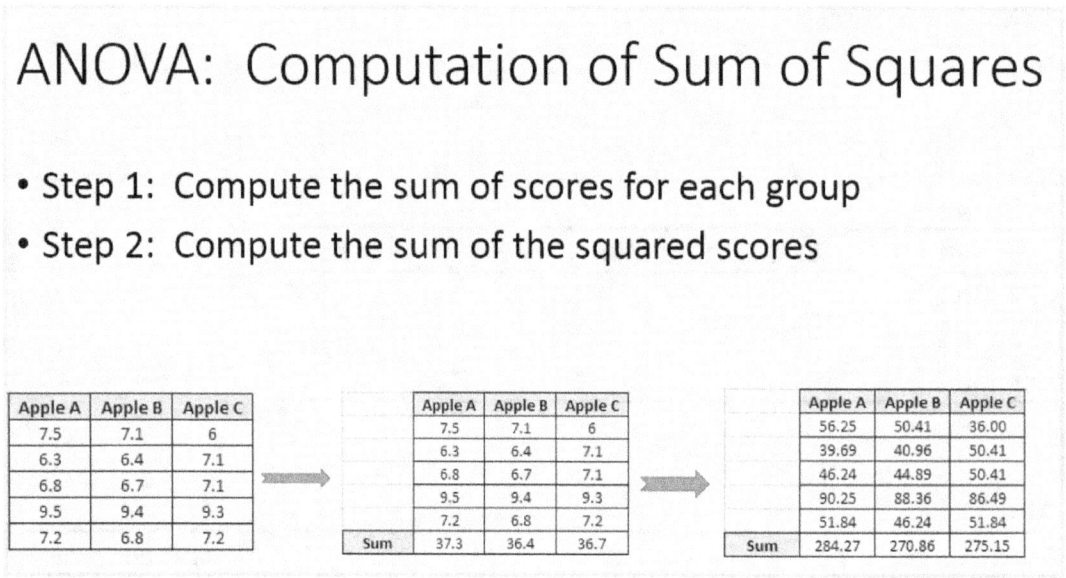

Figure 139: Computation of Sum of Squares Step 1 and 2

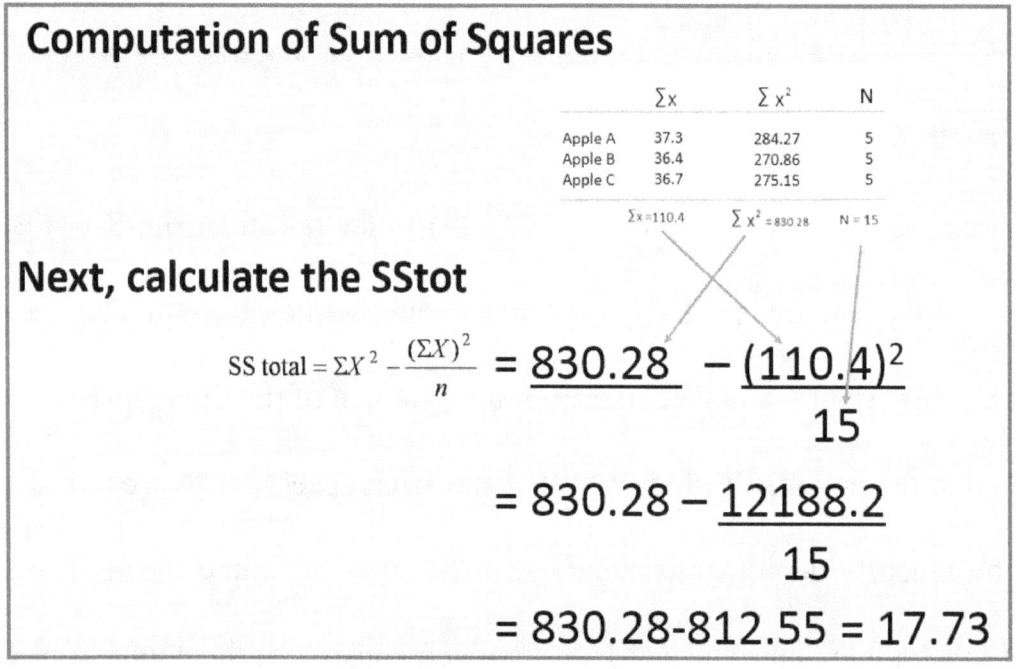

Figure 140: Computation of Sum of Squares and SStot

	$\sum x$	$\sum x^2$	N
Apple A	37.3	284.27	5
Apple B	36.4	270.86	5
Apple C	36.7	275.15	5
	$\sum x = 110.4$	$\sum x^2 = 830.28$	N = 15

Next, calculate the $SS_{between}$

$$SS_b = \sum_g \left[\frac{(\sum X_g)^2}{N_g} \right] - \frac{(\sum X)^2}{N} = \left(\frac{37.3^2}{5} + \frac{36.4^2}{5} + \frac{36.7^2}{5} \right) - \frac{110.4^2}{15}$$

$$= 812.6 - 812.5 = 0.1$$

Figure 141: Calculation of $SS_{between}$

	$\sum x$	$\sum x^2$	N
Apple A	37.3	284.27	5
Apple B	36.4	270.86	5
Apple C	36.7	275.15	5
	$\sum x = 110.4$	$\sum x^2 = 830.28$	N = 15

calculate the SS_{within}

$$SS_w = \sum_g \left[(\sum X_g^2) - \frac{(\sum X_g)^2}{N_g} \right]$$

$$= \left(284.27 - \frac{37.3^2}{5} \right) + \left(270.86 - \frac{36.4^2}{5} \right) + \left(275.15 - \frac{36.7^2}{5} \right)$$

$$= \left(284.27 - \frac{37.3^2}{5} \right) + \left(270.86 - \frac{36.4^2}{5} \right) + \left(275.15 - \frac{36.7^2}{5} \right)$$

$$= 17.652$$

Figure 142: Calculation of SS_{within}

Note that the sum of $SS_{between}$ and $SS_{within} = SS_{tot}$ (0.1+17.6) = 17.7. Refer back to the product that Microsoft® Excel produced and it is the same result. The number of calculations that must be performed manually makes Microsoft® Excel a much better and reliable way to perform ANOVA. The F statistic is computed by dividing the Mean Sqaure$_{between}$ by the Mean Square$_{within}$. The Mean Square$_{between}$ is calculated by dividing the $SS_{between}$ by c-1 (0.1/2). The Mean Square$_{within}$ is calculated by dividing the SS_{within} by n-c (17.6/12). The F statistic is calculated by dividing the Mean Square$_{between}$ by the Mean Square$_{within}$. This is calculated by dividing 0.05 by 1.46. The resulting F statistic is 0.034, which is slightly higher than what Microsoft® Excel calculated, is due to the rounding of numbers done in manual calculations. If a significant difference is suggested by the one-way ANOVA, the next step is to perform a post-hoc (after the fact) analysis. Post-hoc analysis will be discussed in Chapter 18.

Microsoft® Excel also performs the analysis for ANVOA where there may be two effects instead of one (ANOVA: two factor without replication) and where there are two factors in which the data is measured more than once per factor (ANOVA: two factor with replication). The reader is advised to refer to other textbooks for a detailed description of how and when these tests should be used.

Chapter 17 questions

1. Determine any potential difference in means using one way ANOVA on the following dataset. Assume alpha is 0.05 for all tests.

Group 1	Group 2	Group 3
9	5	5
0	2	5
2	6	2
9	11	9
7	7	12
3	9	2
0	6	9
0	10	9

2. Determine any potential difference in means using one way ANOVA on the following dataset.

Group 1	Group 2	Group 3
2	11	9
0	10	2
2	10	2
6	9	10
4	8	5
7	11	7
6	11	11
10	7	7

3. Determine any potential difference in means using one way ANOVA on the following dataset.

Group 1	Group 2	Group 3
7	4	6
7	9	9
5	2	5
1	3	7
3	3	3
4	1	10
3	9	2
0	10	8

4. Determine any potential difference in means using one way ANOVA on the following dataset.

Group 1	Group 2	Group 3
2	8	18
7	10	4
3	3	0
3	2	1
7	1	0
7	10	7
3	1	10
8	0	0

5. Determine any potential difference in means using one way ANOVA on the following dataset.

Group 1	Group 2	Group 3
9	5	8
1	14	20
0	11	6
5	15	17
1	12	14
9	7	30
6	12	22
2	16	12

6.

Group 1	Group 2	Group 3	Group 4	Group 5
14	14	8	4	12
15	15	20	9	24
22	22	6	2	25
6	6	17	3	7
10	10	14	3	7
8	8	30	1	15
13	13	22	9	28
18	18	12	10	17

Chapter 17 Solutions

1. No potential difference noted in means. F < F crit.

Anova: Single Factor

SUMMARY

Groups	Count	Sum	Average	Variance
Column 1	8	30	3.75	15.92857
Column 2	8	56	7	8.571429
Column 3	8	53	6.625	13.41071

ANOVA

Source of Variation	SS	df	MS	F	P-value	F crit
Between Groups	50.58333	2	25.29167	2.001413	0.160109	3.4668
Within Groups	265.375	21	12.6369			
Total	315.9583	23				

2. A potential difference noted in means. F > F crit.

Anova: Single Factor

SUMMARY

Groups	Count	Sum	Average	Variance
Column 1	8	37	4.625	10.55357
Column 2	8	77	9.625	2.267857
Column 3	8	53	6.625	11.69643

ANOVA

Source of Variation	SS	df	MS	F	P-value	F crit
Between Groups	101.3333	2	50.66667	6.199563	0.007657	3.4668
Within Groups	171.625	21	8.172619			
Total	272.9583	23				

3. No potential difference noted in means. F < F crit.

Anova: Single Factor

SUMMARY

Groups	Count	Sum	Average	Variance
Column 1	8	30	3.75	6.5
Column 2	8	41	5.125	12.98214
Column 3	8	50	6.25	7.928571

ANOVA

Source of Variation	SS	df	MS	F	P-value	F crit
Between Groups	25.08333	2	12.54167	1.372638	0.275259	3.4668
Within Groups	191.875	21	9.136905			
Total	216.9583	23				

4. No potential difference noted in means. F < F crit.

Anova: Single Factor

SUMMARY

Groups	Count	Sum	Average	Variance
Column 1	8	40	5	6
Column 2	8	35	4.375	17.98214
Column 3	8	40	5	41.42857

ANOVA

Source of Variation	SS	df	MS	F	P-value	F crit
Between Groups	2.083333	2	1.041667	0.047775	0.953452	3.4668
Within Groups	457.875	21	21.80357			
Total	459.9583	23				

5. A potential difference noted in means. F > F crit.

Anova: Single Factor

SUMMARY

Groups	Count	Sum	Average	Variance
Column 1	8	33	4.125	13.26786
Column 2	8	92	11.5	14.57143
Column 3	8	129	16.125	61.83929

ANOVA

Source of Variation	SS	df	MS	F	P-value	F crit
Between Groups	586.0833	2	293.0417	9.803067	0.000984	3.4668
Within Groups	627.75	21	29.89286			
Total	1213.833	23				

6. A potential difference noted in means. F > F crit

Anova: Single Factor

SUMMARY

Groups	Count	Sum	Average	Variance
Column 1	8	106	13.25	27.64286
Column 2	8	106	13.25	27.64286
Column 3	8	129	16.125	61.83929
Column 4	8	41	5.125	12.98214
Column 5	8	135	16.875	66.125

ANOVA

Source of Variation	SS	df	MS	F	P-value	F crit
Between Groups	695.15	4	173.7875	4.42811	0.005313	2.641465
Within Groups	1373.625	35	39.24643			
Total	2068.775	39				

CHAPTER 18: POST-HOC ANALYSIS

After recognizing there is a statistical difference in means, post ANOVA, the next step in the process is to determine which means are actually different. Consider the example of the weights of oranges grown in three different states (Florida, Texas, and California). The ANOVA reveals a statistically significant difference in means. Is Florida different from Texas or California….or both? Is California different from Texas or Florida….or both? Is California different from Texas or Florida…..or both? The issue of discriminating between means was not an issue when only two groups were analyzed because there could be only two answers, that being different or not different.

Two specific post hoc tests will be discussed in this Chapter. There are downloads available for Microsoft® Excel that allow completion of both of the post hoc tests discussed. However, the simplicity of tests allows for simple manual calculations. This chapter will cover the manual calculations. The first is the Tukey's Honestly Significant Difference test (HSD) and the second is the Fisher's Least Significant Difference test (LSD). It should be noted that there are many post-hoc tests available and the reader is advised to research and/or consult with a professional statistician to find the appropriate post-hoc test.

The Turkey's HSD is an easy test to perform. There are three conditions which must occur for Tukey's HSD to be valid. First, the observations must be independent within and between groups. Second, the variance within the groups must not be different when compared to other groups. Finally, the data must be normally distributed. The HSD works best for equal sample sizes. There are methods to perform a HSD on unequal sample sizes and the reader is encouraged to review these in other textbooks. The formula for the HSD is as follows in Figure 143:

$$HSD = q \sqrt{MSE/n}$$

Figure 143: Tukey's HSD Formula

In the HSD formula, q is the critical value for Q scores for Turkey's method. There are many online resources and the reader is encouraged to refer to these sources to obtain Q scores for Turkey's method. An example of such a source can be found at the following website - http://www.stat.wisc.edu/courses/st571-ane/tables/tableQ.pdf. The MSE is the Within Groups Mean Square, referred to earlier as MS_{within}. The n is the number of samples within each group.

Consider the following dataset comparing 3 groups. See Figure 144:

A	B	C
2	3	9
3	3	8
2	3	7
3	1	9
2	2	8
3	2	7

Figure 144: Tukey HSD Dataset

This dataset yielded the following ANOVA table if alpha is 0.05. See Figure 145:

Anova: Single Factor						
SUMMARY						
Groups	Count	Sum	Average	Variance		
Column 1	6	15	2.5	0.3		
Column 2	6	14	2.333333	0.666667		
Column 3	6	48	8	0.8		
ANOVA						
Source of Variation	SS	df	MS	F	P-value	F crit
Between Groups	124.7778	2	62.38889	105.9434	1.42E-09	3.68232
Within Groups	8.833333	15	0.588889			
Total	133.6111	17				

Figure 145: ANVOA Output

Using the previously mentioned equation, the HSD can be calculated. The difference between two sample means is considered significant if it exceeds HSD. Recall the HSD equation is:

$$HSD = q \sqrt{MSE / n}$$

From a table of q values, q (3 groups, df of 15), is found to be 3.67. Tueky's HSD is calculated as shown in Figure 146:

$$HSD = 3.67\sqrt{\frac{0.589}{6}} = 3.67 \times 0.313 = 1.15$$

Figure 146: Calculation of Tukey HSD

In other words, there would need to be a difference of at least 1.15 to be considered significant. In this example, there would be no difference between Group A and B, but there would be a statistically significant difference between Group C and Group A and between Group C and Group B.

The other post hoc test that will be covered in this chapter is the Fisher's Least Significant Difference (LSD) test. The LSD, like the Tukey HSD, is easy to compute and is used to make comparisons between means. Although not considered as statistically sound as the HSD, it does have the advantage of handling unequal sample sizes without additional steps. The LSD between two sample means is calculated with the following equation (Figure 147):

$$LSD = \frac{\overline{X_1} - \overline{X_2}}{\sqrt{MS_W\left(\frac{1}{N_1} + \frac{1}{N_2}\right)}}$$

Figure 147: Fisher's Least Significant Difference (LSD) Test

Consider the following data and ANOVA table (Figures 148 and 149).

A	B	C
2	3	9
3	2	8
1	3	9
3	1	9
2	2	6
3	3	7

Figure 148: Fisher's LSD Data

Anova: Single Factor

SUMMARY

Groups	Count	Sum	Average	Variance
Column 1	6	14	2.333333	0.666667
Column 2	6	14	2.333333	0.666667
Column 3	6	48	8	1.6

ANOVA

Source of Variation	SS	df	MS	F	P-value	F crit
Between Groups	128.4444	2	64.22222	65.68182	3.8E-08	3.68232
Within Groups	14.66667	15	0.977778			
Total	143.1111	17				

Figure 149: ANOVA Output

Each pair of data must be compared individually. It can be seen from the ANOVA table that F is very large, so the conclusion is there is significant evidence there is a difference in means. There will be three comparisons made. See Figure 150:

DIFFERENCE A – B	$\dfrac{2.33-2.33}{\sqrt{0.978\left(\frac{1}{6}+\frac{1}{6}\right)}}$	$=\dfrac{0}{0.571}$	$= 0.00$
DIFFERENCE C – A	$\dfrac{8.0-2.33}{\sqrt{0.978\left(\frac{1}{6}+\frac{1}{6}\right)}}$	$=\dfrac{5.67}{0.571}$	$= 9.94$
DIFFERENCE C - B	$\dfrac{8.0-2.33}{\sqrt{0.978\left(\frac{1}{6}+\frac{1}{6}\right)}}$	$=\dfrac{5.67}{0.571}$	$= 9.94$

Figure 150: Fisher's LSD Comparisons

Refer to Table 2, the t critical is one way, assume alpha is 0.05. Since it is only a "difference" that is under consideration, a two tailed test is recommended. The t crit for df of 15 (18-3) and alpha of 0.05 is 2.131. 9.94 is greater than 2.131, so the conclusion is evidence suggests a difference exists.

Chapter 18 Problems – Assume alpha is 0.05 for any test.

1. For the following dataset, perform one way ANOVA and calculate the HSD and LSD for each if necessary..

GROUP 1	GROUP 2	GROUP 3
1	2	4
2	3	3
3	2	2
4	3	4
5	2	3
6	3	2

2. For the following dataset, perform one way ANOVA and calculate the HSD and LSD for each if necessary.

GROUP 1	GROUP 2	GROUP 3
1	2	3
4	5	6
7	8	9
10	12	14
1	2	3

3. For the following dataset, perform one way ANOVA and calculate the HSD and LSD for each if necessary.

GROUP 1	GROUP 2	GROUP 3
4	6	4
3	7	5
2	5	4
1	6	5
2	7	4
3	5	5
4	6	3

Chapter 18 Solutions

1. The ANOVA table does NOT indicate there is a probable difference in means $F < F$ crit. Therefore, it is not necessary to perform HSD or LSD.

Anova: Single Factor

SUMMARY

Groups	Count	Sum	Average	Variance
Column 1	6	21	3.5	3.5
Column 2	6	15	2.5	0.3
Column 3	6	18	3	0.8

ANOVA

Source of Variation	SS	df	MS	F	P-value	F crit
Between Groups	3	2	1.5	0.978261	0.398711	3.68232
Within Groups	23	15	1.533333			
Total	26	17				

2. The ANOVA table does NOT indicate there is a probable difference in means $F < F$ crit. Therefore, it is not necessary to perform HSD or LSD.

Anova: Single Factor

SUMMARY

Groups	Count	Sum	Average	Variance
Column 1	5	23	4.6	15.3
Column 2	5	29	5.8	18.2
Column 3	5	35	7	21.5

ANOVA

Source of Variation	SS	df	MS	F	P-value	F crit
Between Groups	14.4	2	7.2	0.392727	0.683582	3.885294
Within Groups	220	12	18.33333			
Total	234.4	14				

3. The ANOVA table does indicate there is a probable difference in means $F > F$ crit.

Anova: Single Factor

SUMMARY

Groups	Count	Sum	Average	Variance
Column 1	7	19	2.714286	1.238095
Column 2	7	42	6	0.666667
Column 3	7	30	4.285714	0.571429

ANOVA

Source of Variation	SS	df	MS	F	P-value	F crit
Between Groups	37.80952	2	18.90476	22.90385	1.13E-05	3.554557
Within Groups	14.85714	18	0.825397			
Total	52.66667	20				

HSD Test:

$$HSD = q \sqrt{MSE/n}$$

From a table of q values, q (3 groups, df of 18), is found to be 3.61

$$HSD = 3.67 \sqrt{\frac{0.825}{7}} = 3.67 \times 0.343 = 1.26$$

The conclusion using the HSD is that any difference of >1.26 is considered significant. All groups are significantly different from each other.

LSD Test:

Difference A-B $\dfrac{6 - 2.7}{\sqrt{0.82\,(1/7+1/7)}} = \dfrac{3.3}{0.234} = 14.1$

Difference A-C $\dfrac{4.3 - 2.7}{\sqrt{0.82\,(1/7+1/7)}} = \dfrac{1.6}{0.234} = 6.8$

Difference B-C $\dfrac{6 - 4.3}{\sqrt{0.82\,(1/7+1/7)}} = \dfrac{1.7}{0.234} = 7.3$

From Table 2, the t critical for df = 18, alpha 0.05 is 2.101. Each difference is greater than 2.1, thus all are considered different.

CHAPTER 19: TESTS FOR DIFFERENCES IN PROPORTIONS

One of the most useful statistical tools in the real world is the ability to discriminate between two proportions. Full textbooks have been written on this topic alone. This chapter will very briefly cover two commonly used tests, the Chi Square and Fisher's exact test.

Tests to determine differences in proportions examine the expected and observed outcomes that are divided into categories. In other words, the results can be in only one category. The frequency of outcomes can be displayed in a contingency table. An example of a contingency table is shown below, a poll of football fans from Columbus Mississippi. See Figure 151:

	Football Team		
ROOT FOR?	MSU	Ole Miss	Total
Yes	130	70	200
No	100	160	260
Total	230	230	460

Figure 151: Ole Miss and MSU Fans

This example is of a 2x2 contingency table because there are four outcomes presented (2x2). Contingency tables can contain any number of rows and columns.

In problems associated with proportions, the null hypothesis is there is no difference in the proportion. The alternate hypothesis is there are differences in the proportion.

For the chi square test, the chi square (χ^2) statistic must be calculated. The formula for the χ^2 statistic is as follows in Figure 152:

$$\chi^2 = \sum \frac{(\text{observed} - \text{expected})^2}{\text{expected}}$$

Figure 152: Chi Square ($\chi 2$) Statistic

For the χ^2 work correctly, there are a couple of assumptions. Each cell in the contingency table must be great than five and there must be no ordering of the data. This test can be calculated manually and there are numerous free websites that can quickly perform the analysis (see http://graphpad.com/quickcalcs/contingency1.cfm). Microsoft® Excel also perform the analysis. Before advancing to use Microsoft® Excel, the data must be organized in proper fashion. Consider the following data to see if there is a difference in number of hours of television watched per week for students who have > and < 3.0 GPA at a suburban university. See Figure 153:

		> 3.0 GPA	< 3.0 GPA
# hours T.V Watched	< 10 hours	26	11
	> 10 hours	25	48

Figure 153: Hours TV Watched vs. GPA

The first step is to calculate the totals by row and column. See Figure 154:

		> 3.0 GPA	< 3.0 GPA	Total
# hours T.V Watched	< 10 hours	26	11	37
	> 10 hours	25	48	73
	Total	51	59	110

Figure 154: Total Rows and Columns

Next, the expected values are needed for each cell. The expected values are calculated by multiplying the row by the column for each associated cell and dividing that number by the total number of observations. The calculations in each cell are shown as follows in Figure 155:

		> 3.0 GPA	< 3.0 GPA	Total
# hours T.V Watched	< 10 hours	(51*37)/100	(59*37)/110	37
	> 10 hours	(51*73)/110	(59*73)/110	73
	Total	51	59	110

Figure 155: Expected Values for Cell I

The expected numbers, after the calculations have been made, are shown in Figure 156:

		> 3.0 GPA	< 3.0 GPA	Total
# hours T.V Watched	< 10 hours	(51*37)/110	(59*37)/110	37
	> 10 hours	(51*73)/110	(59*73)/110	73
	Total	51	59	110

Figure 156: Expected Values for Cell II

The data from the observed and expected frequencies should be entered and analyzed into Microsoft® Excel as follows in Figure 157:

1. Arrange the data so it is easily seen on one screen.

Observed		> 3.0 GPA	< 3.0 GPA	Total
# hours T.V Watched	< 10 hours	26	11	37
	> 10 hours	25	48	73
	Total	51	59	110

Expected		> 3.0 GPA	< 3.0 GPA	Tot
# hours T.V Watched	< 10 hours	17.15454545	19.84545	3
	> 10 hours	33.84545455	39.15455	7
	Total	51	59	11

Figure 157: Arranging Data - Fisher's Exact Test

2. From the Formulas tab, select insert function and then select "chisqtest". Older versions of Microsoft® Excel may require the use of "chitest". The dialogue box showing chiqtest will appear. See Figure 158:

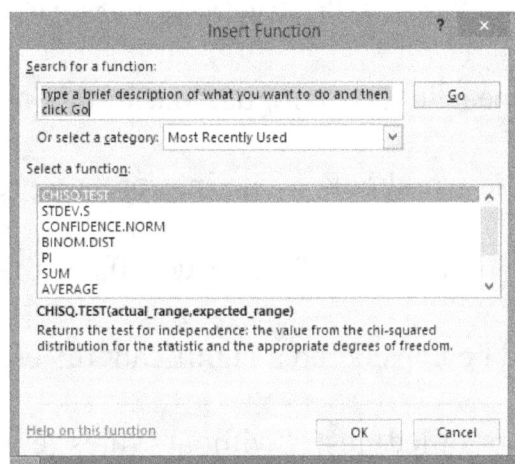

Figure 158: Chisqtest Dialogue Box

Next, input the actual and highlighted range into the dialogue box. It should appear as follows in Figure 159:

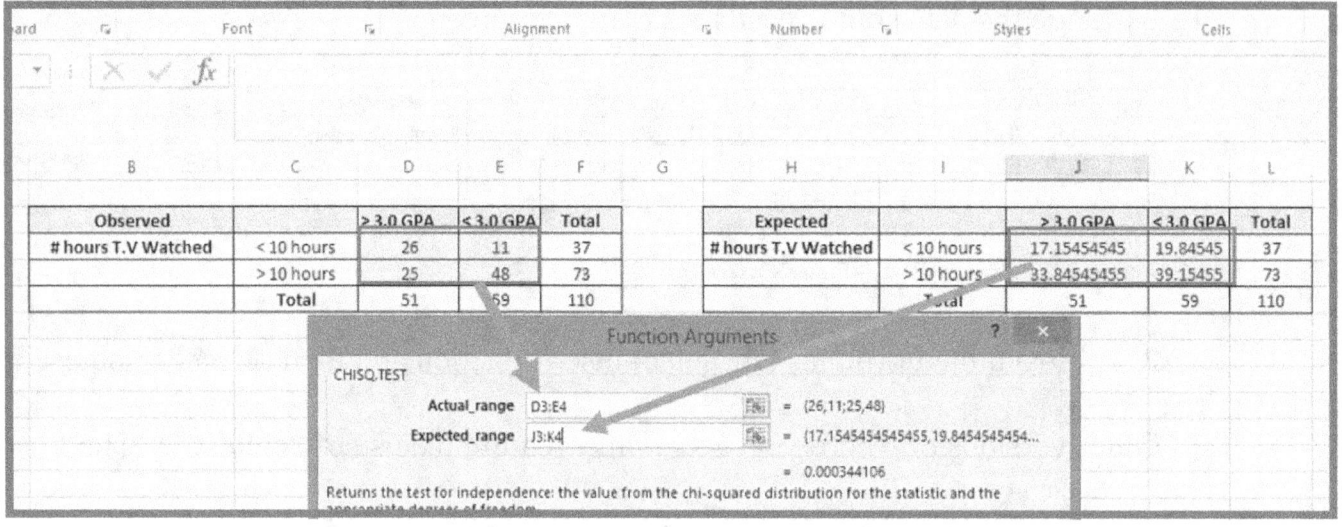

Figure 159: Highlighted Range in Chitest Dialogue Box

The result will appear in the dialogue box, in this case it is 0.0003. In addition, it will also appear on the workbook. The interpretation of this result depends on the alpha set forth in the hypothesis. If alpha was 0.05, then this result is far less than 0.05, therefore it would be considered a significant difference. Please note that excel calculates χ^2 as a two tailed test "without yates correction". As previously mentioned, there are many online resources to perform χ^2. One advantage to these resources is not needing to input the expected values as that is automatically calculated.

If any values in the contingency table are five or less, it is inappropriate to use χ^2. In these circumstances, the Fisher's Exact Test should be used. Microsoft® Excel does not have built in capability to perform Fisher's Exact Test. There are commercial software programs that will allow for this analysis along with user developed Microsoft® Excel templates. For example, a template can be downloaded from the following website:

- http://commfaculty.fullerton.edu/jreinard/excelexcel.htm.

The quickest way is to use Fisher's exact test is via one of the many free websites. Graphpad software has a really good site that is very user friendly. The site can be found at:

http://graphpad.com/quickcalcs/contingency1.cfm

The following is an example from this website. See Figure 160:

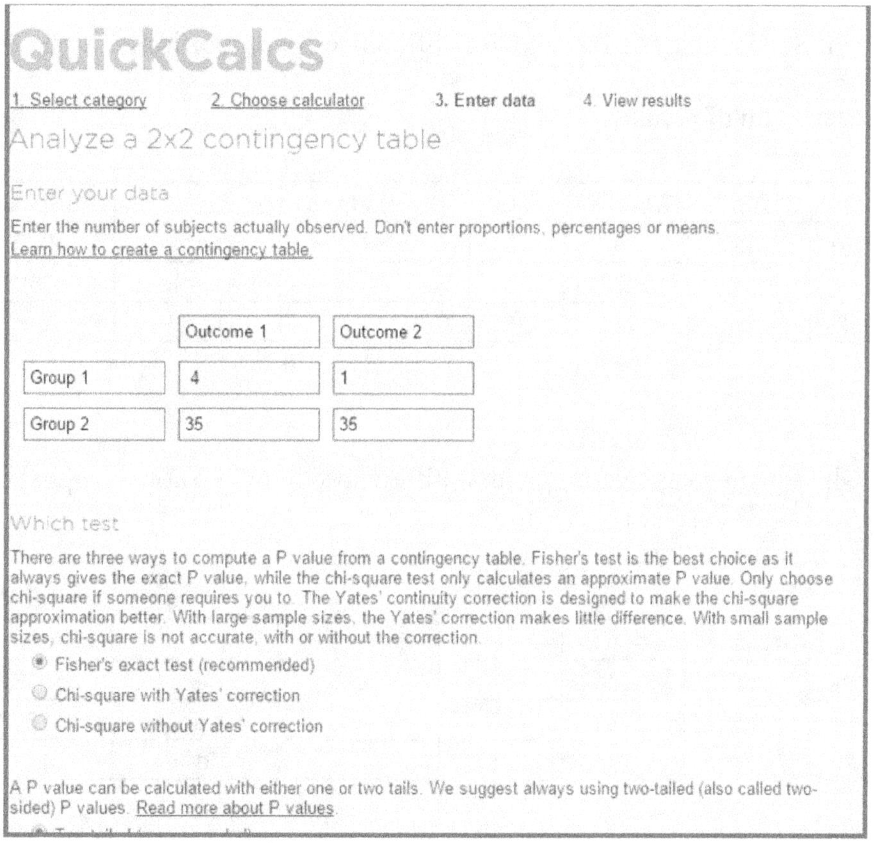

Figure 160: GraphPad 2x2 Contingency Table

Chapter 19 Problems:

1. Perform Chi Square analysis for potential differences in proportions. Alpha is 0.05. Give the answer in form of p value.

Observed		Group A	Group B	Total
XXXX	Variable 1	17	35	52
	Variable 2	28	59	87
	Total	45	94	139

Expected		Group A	Group B	Total
XXXX	Variable 1	16.83453	35.16547	52
	Variable 2	28.16547	58.83453	87
	Total	45	94	139

2. Perform Chi Square analysis for potential differences in proportions. Alpha is 0.05. Give the answer in form of p value.

Observed		Group A	Group B	Total
XXXX	Variable 1	25	45	70
	Variable 2	11	56	67
	Total	36	101	137

Expected		Group A	Group B	Total
XXXX	Variable 1	18.39416	51.60584	70
	Variable 2	17.60584	49.39416	67
	Total	36	101	137

3. Perform Chi Square analysis for potential differences in proportions. Alpha is 0.05. Give the answer in form of p value.

Observed		Group A	Group B	Total
XXXX	Variable 1	45	13	58
	Variable 2	33	8	41
	Total	78	21	99

Expected		Group A	Group B	Total
XXXX	Variable 1	45.69697	12.30303	58
	Variable 2	32.30303	8.69697	41
	Total	78	21	99

4. Perform Chi Square analysis for potential differences in proportions. Alpha is 0.05. Give the answer in form of p value.

Observed		Group A	Group B	Total
XXXX	Variable 1	66	32	98
	Variable 2	38	41	79
	Total	104	73	177

Expected		Group A	Group B	Total
XXXX	Variable 1	57.58192	40.41808	98
	Variable 2	46.41808	32.58192	79
	Total	104	73	177

5. Perform Chi Square analysis for potential differences in proportions. Alpha is 0.05. Give the answer in form of p value.

Observed		Group A	Group B	Total		Expected		Group A	Group B	Total
XXXX	Variable 1	20	21	41		XXXX	Variable 1	25.38095	15.61905	41
	Variable 2	32	11	43			Variable 2	26.61905	16.38095	43
	Total	52	32	84			Total	52	32	84

6. Perform Chi Square analysis for potential differences in proportions. Alpha is 0.05. Give the answer in form of p value.

Observed		Group A	Group B	Total		Expected		Group A	Group B	Total
XXXX	Variable 1	19	62	81		XXXX	Variable 1	24.3	56.7	81
	Variable 2	14	15	29			Variable 2	8.7	20.3	29
	Total	33	77	110			Total	33	77	110

7. Perform Chi Square analysis for potential differences in proportions. Alpha is 0.05. Give the answer in form of p value.

Observed		Group A	Group B	Total		Expected		Group A	Group B	Total
XXXX	Variable 1	31	33	64		XXXX	Variable 1	32.66207	31.33793	64
	Variable 2	43	38	81			Variable 2	41.33793	39.66207	81
	Total	74	71	145			Total	74	71	145

8. Perform Fisher's exact analysis for potential differences in proportions. Alpha is 0.05. Give the answer in form of p value.

Observed		Group A	Group B	Total
XXXX	Variable 1	31	33	64
	Variable 2	17	4	21
	Total	48	37	85

9. Perform Fisher's exact analysis for potential differences in proportions. Alpha is 0.05. Give the answer in form of p value.

Observed		Group A	Group B	Total
XXXX	Variable 1	14	19	33
	Variable 2	1	3	4
	Total	15	22	37

10. Perform Fisher's exact analysis for potential differences in proportions. Alpha is 0.05. Give the answer in form of p value.

Observed		Group A	Group B	Total
XXXX	Variable 1	0	9	9
	Variable 2	3	49	52
	Total	3	58	61

Chapter 19 Solutions:

1. The two-tailed P value equals 0.9506.

2. The two-tailed P value equals 0.0103.

3. The two-tailed P value equals 0.7279.

4. The two-tailed P value equals 0.0097.

5. The two-tailed P value equals 0.0156.

6. The two-tailed P value equals 0.0123.

7. The two-tailed P value equals 0.5782.

8. The two-tailed P value equals 0.0112.

9. The two-tailed P value equals 0.6328

10. The two-tailed P value equals 1.0000

CHAPTER 20: CORRELATION AND REGRESSION

This chapter will very briefly consider the concepts of correlation and regression. The major objective of this chapter is to introduce Microsoft® Excel's ability to easily produce correlation and regression products.

Correlation is a fairly simple concept. It is the degree of relationship between variables. For example, a researcher many be interested in the degree of relationship between height and weight.

In general terms, there are three classes of correlation: positive, negative, and zero. See Figures 161, 162, and 163:

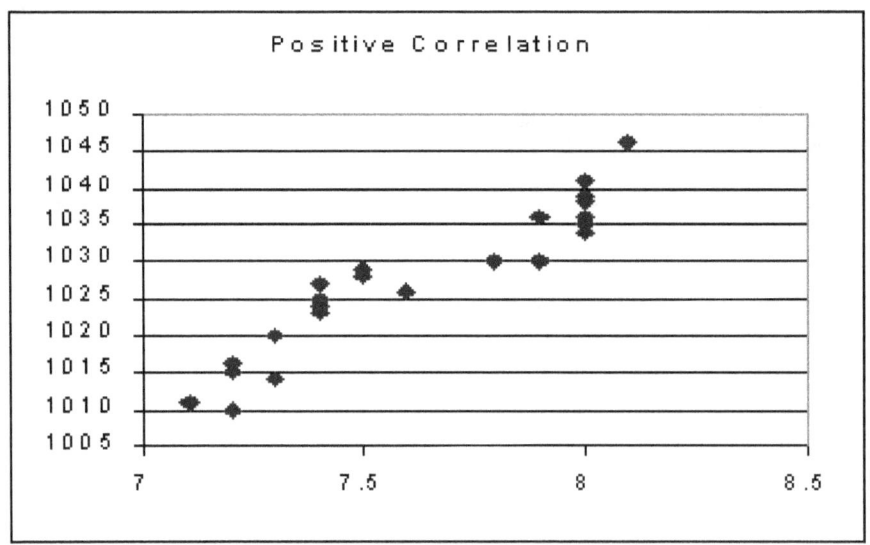

Figure 161: Positive Correlation

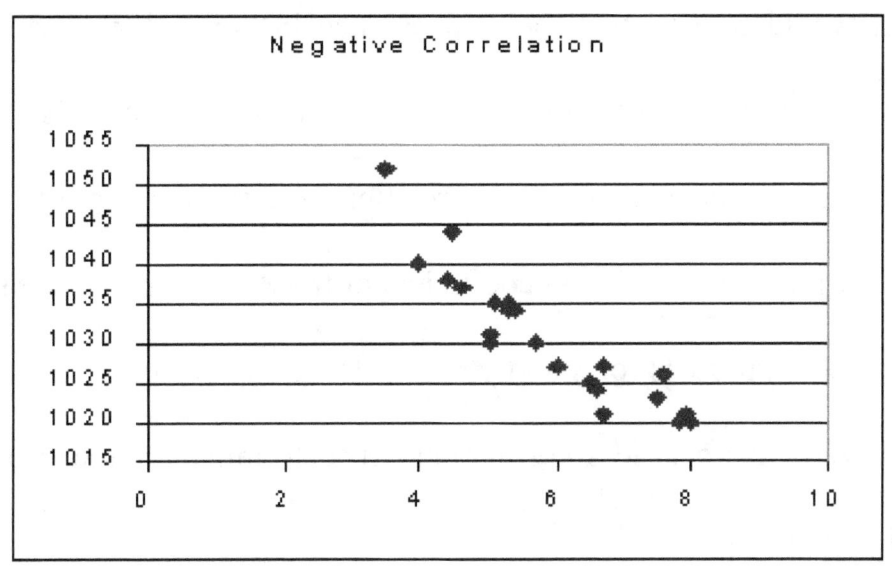

Figure 162: Negative Correlation

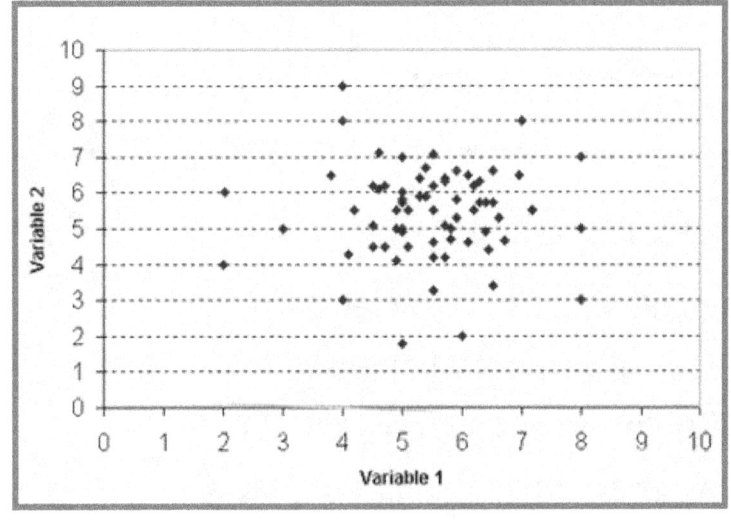

Figure 163: Zero Correlation

One important facet of correlation is that correlation <u>does not</u> equal causation. For example, one may see a young woman with a small child in the park every afternoon. It would be true that their appearance in the park is correlated, but it would be an assumption to assume the young woman is the child's mother. She could be a relative, a nanny, etc... Correlation is easy to quantify by using the Pearson product-moment correlation coefficient or the Pearson r. The Pearson r is the mean of the z-score products for x and y pairs. The range of scores is from -1 to +1. The closer to zero, there is less correlation. The formula for the Pearson r is as follows in Figure 164:

$$r = \frac{\sum xy}{\sqrt{\sum x^2 \sum y^2}}$$

Figure 164: Pearson r Formula

The interpretation of the Pearson r varies from source to source. Figure 165 is an example of how the Pearson r may be interpreted:

Interpretation
- -1.0 to -0.7 strong negative association.
- -0.7 to -0.3 weak negative association.
- -0.3 to +0.3 little or no association.
- +0.3 to +0.7 weak positive association.
- +0.7 to +1.0 strong positive association.

- Source - http://www.childrens-mercy.org/stats/definitions/correlation.htm

Figure 165: Pearson r Interpretation

The Pearson r is a descriptive statistic. We can, however, use linear regression in a predictive manner. As a quick review, the general equation for a straight line is as shown in Figure 166:

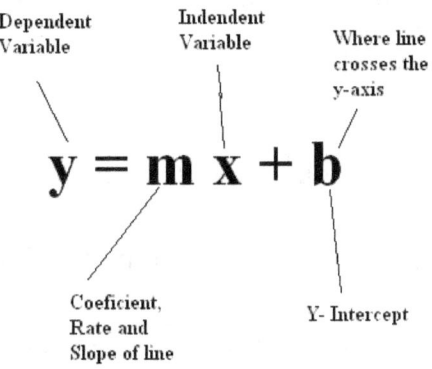

Figure 166: General Equation for a Straight Line

The regression equation is for the straight line that best describes the relationship between the variables. This is called least squares line. See Figure 167:

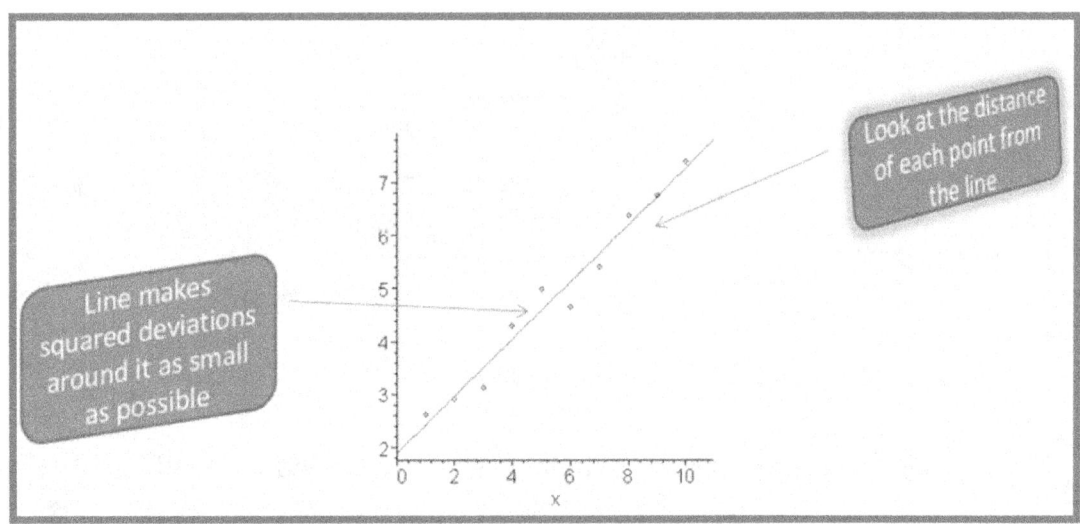

Figure 167: Least Sqaures Line

The product of a linear regression is the formula for a straight line whereby predictions can be made. For example, the product would be y=13.5x + 37. In this situation, if one knows either y or x, then the other can be solved for. Note that linear regression creates the best fitting line thru the data. The less variability in the data, the more accurate the line is.

From linear regression, another product can be generated, the r squared, also known as the coefficient of determination. The coefficient of determination explains the amount of variability in the correlation. For example, an r of 0.9 has an r^2 of 0.81. This means that 81 percent of the overall variability is due to the relationship between the variables. The math is a little more complex, but it can also be calculated by squaring the Pearson r. Microsoft® Excel can also perform this analysis.

Consider the following data (Figure 168):

X	Y
2	4
3	5.8
2	4.1
5	10.2
4	8
11	20.9

Figure 168: r Squared Dataset

In Microsoft® Excel, select data analysis, regression, highlight the field in which data depicting the dependent and independent variables appear. See Figures 169 and 170:

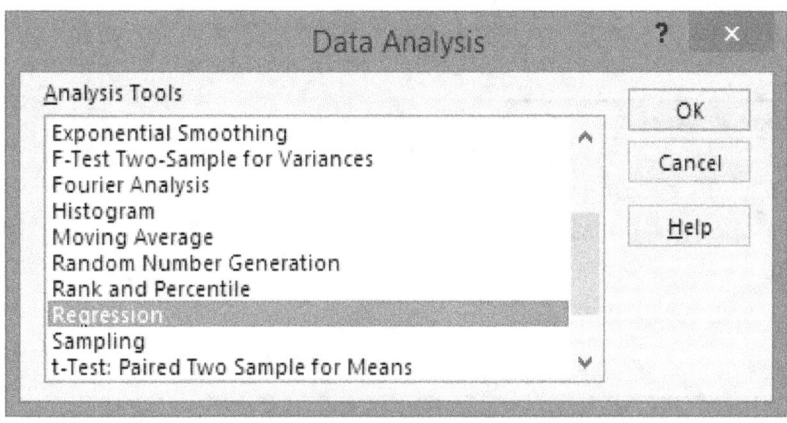

Figure 169: r Squared Setup I

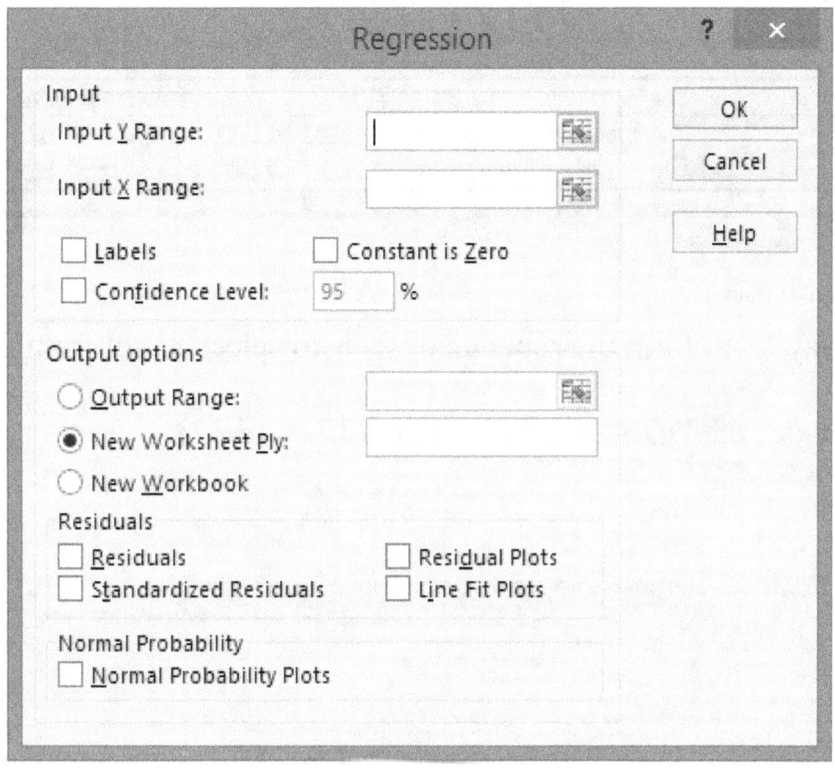

Figure 170: r Squared Setup II

After iputting the data and pressing "OK", the following output should appear as shown in Figure 171. Note both the R and R Sqaure are produced.

SUMMARY OUTPUT

Regression Statistics	
Multiple R	0.999307704
R Square	0.998615887
Adjusted R Square	0.998269859
Standard Error	0.141055378
Observations	6

ANOVA

	df	SS	MS	F	gnificance F
Regression	1	57.42041	57.42041	2885.938	7.19E-07
Residual	4	0.079586	0.019897		
Total	5	57.5			

	Coefficients	andard Err	t Stat	P-value	Lower 95%	Upper 95%	ower 95.0%	pper 95.0%
Intercept	-0.196422711	0.104684	-1.87633	0.133847	-0.48707	0.094228	-0.48707	0.094228
X Variable 1	0.531670496	0.009897	53.72093	7.19E-07	0.504192	0.559149	0.504192	0.559149

Figure 171: r Squared Output

Another way to perform the same analysis is to select correl from functions and get r. This provides the Pearson r. See Figures 172 and 173:

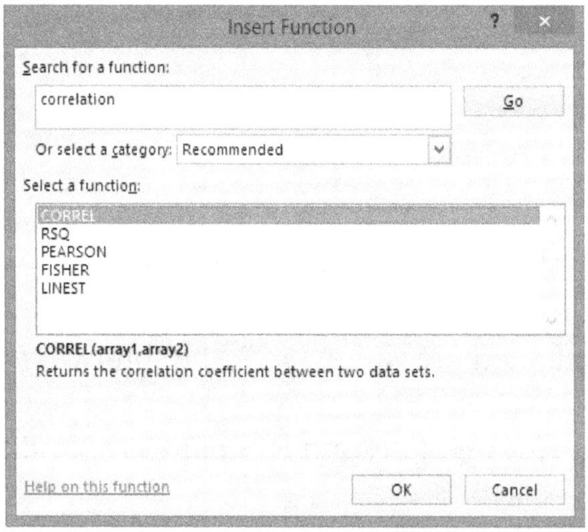

Figure 172: r Squared Using CORREL I

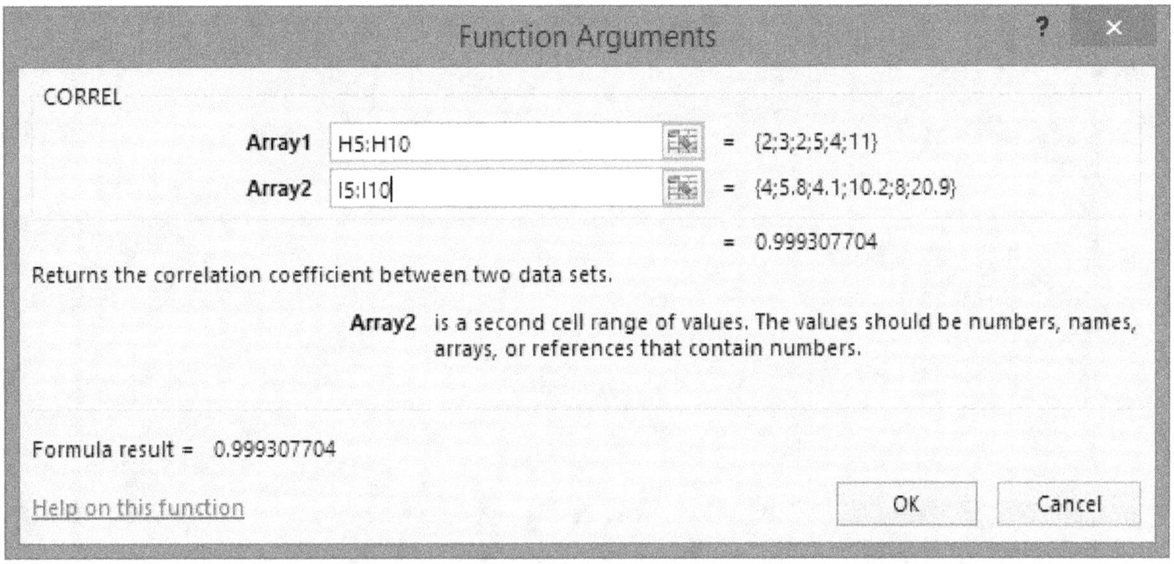

Figure 173: r Squared Using CORREL II

Notice that this r is exactly the same as the r produced in regression. To get r squared, simply square the r. In other words, multiply r by itself.

There is one other way to obtain r and r square. Use the data to make a scatterplot (Insert, Chart, Scatterplot). It should look like Figure 174:

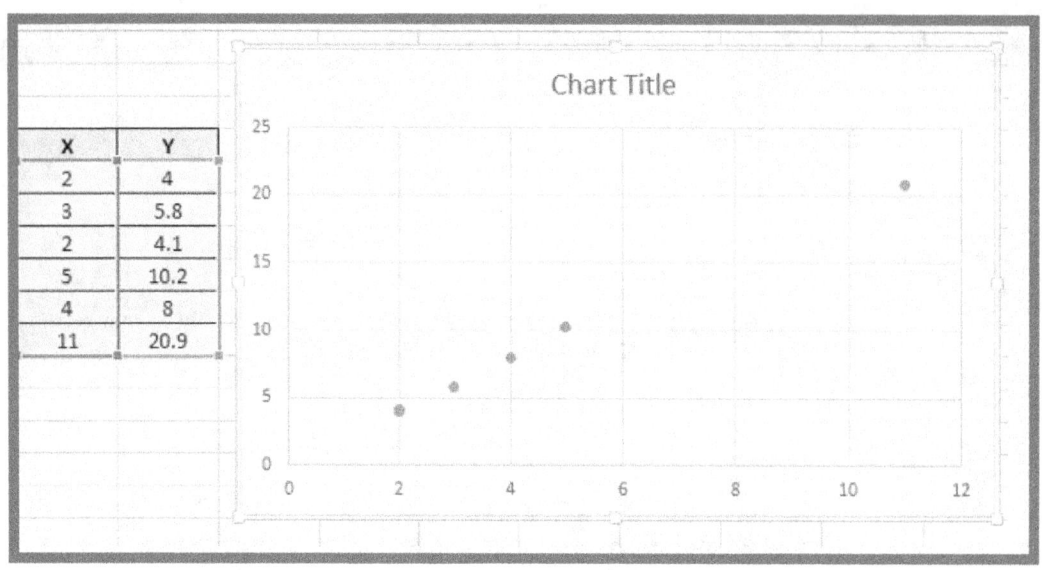

Figure 174: Scatterplot to Obtain r Squared

Next, click on any data point until what appears to be an "X" is seen on all data points. The data on the workbook will also become highlighted and there "series" of data will be highlighted. This can be seen in Figures 175 and 176.:

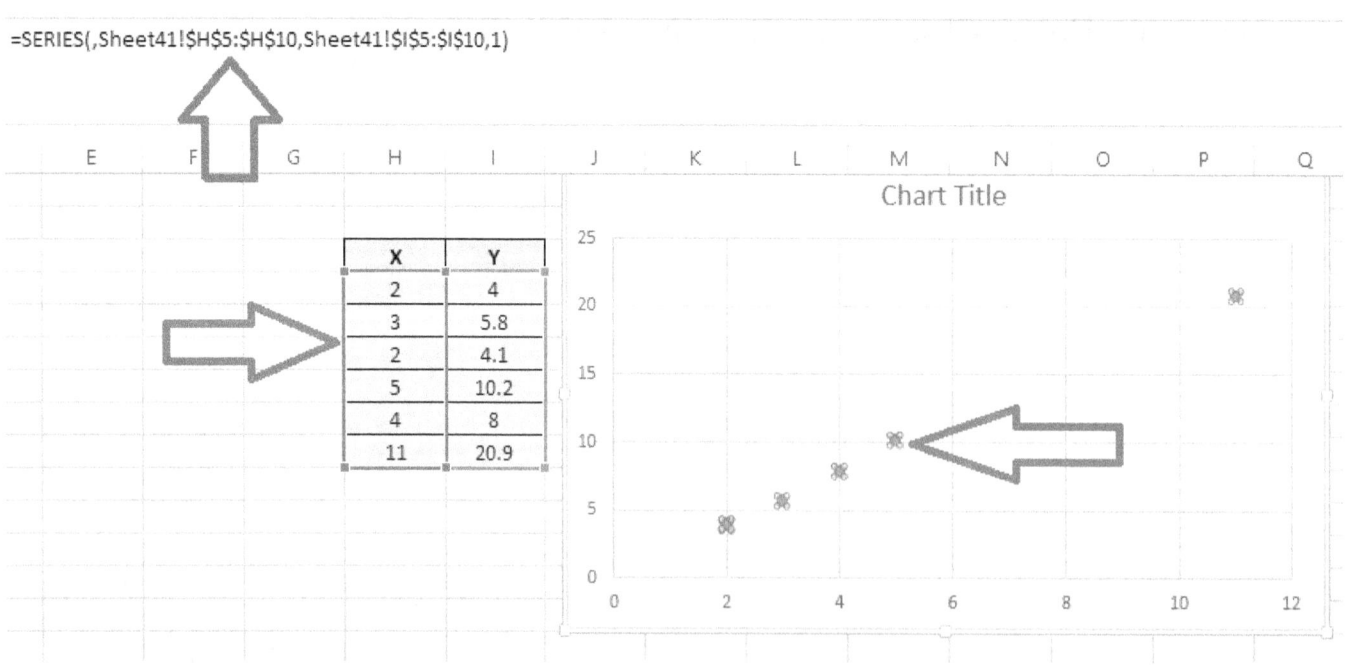

Figure 175: r Squared Using Scatterplot I

Next, right click and select add trendline. See Figure 175:

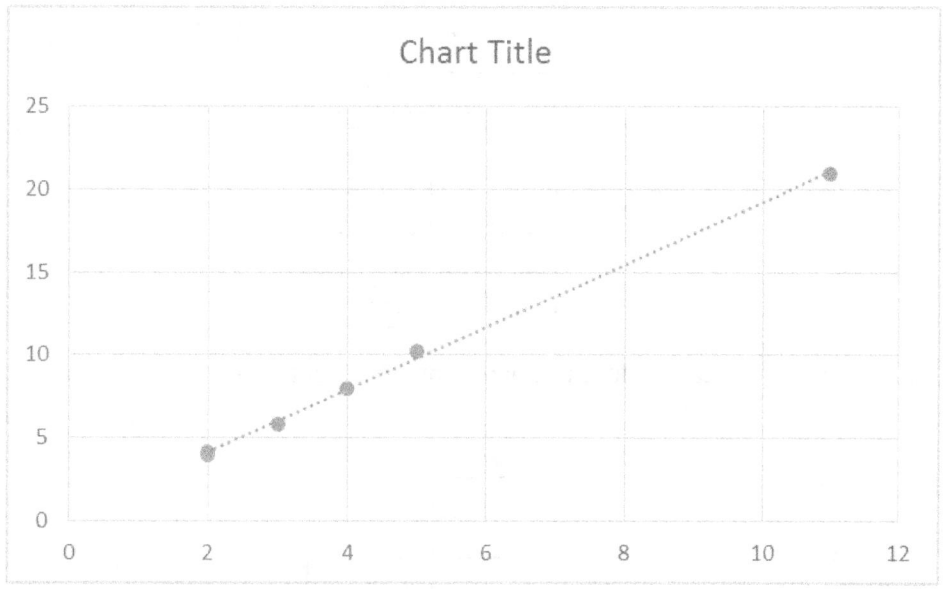

Figure 176: r Squared Using Scatterplot II

A trendline will now appear. Next, right-click on trendline and select "format trendline". At the bottom of the "format trendline" choices will be selections for "Display Equation on Chart" and "Display R-Squared Value on Chart". Select both of these and the equation for the line and r squared should be displayed on the scatterplot. To obtain r, simply take the square root of r squared.

Chapter 20 Questions:

1. For the following data, calculate r, r squared, and the equation for the line.

X	Y
2	6
3	5.8
2	5.4
5	13.5
4	9
11	23.6

2. For the following data, calculate r, r squared, and the equation for the line.

X	Y
3	7
4	7
5	10
6	4
7	17
8	23.6

3. For the following data, calculate r, r squared, and the equation for the line.

X	Y
5	10.1
6	12.2
7	13.3
8	16.4
9	18.5
10	16.5

4. For the following data, calculate r, r squared, and the equation for the line.

X	Y
11	9
11	25
9	13
6	22
1	20
12	25

Chapter 20 Solutions:

1. r=0.9855

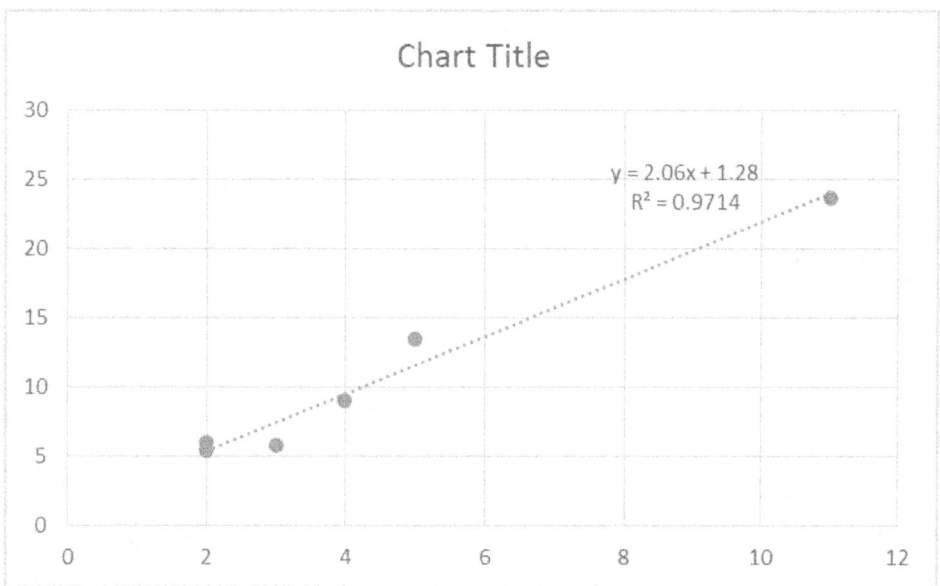

2. r = <0.001 : excel gives 0.0000000099625062625

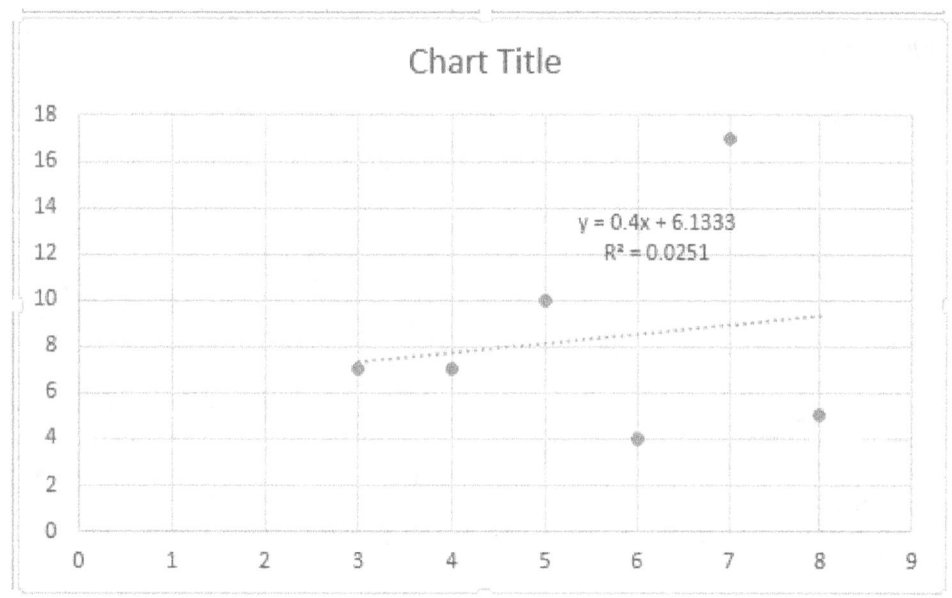

3. r = 0.914

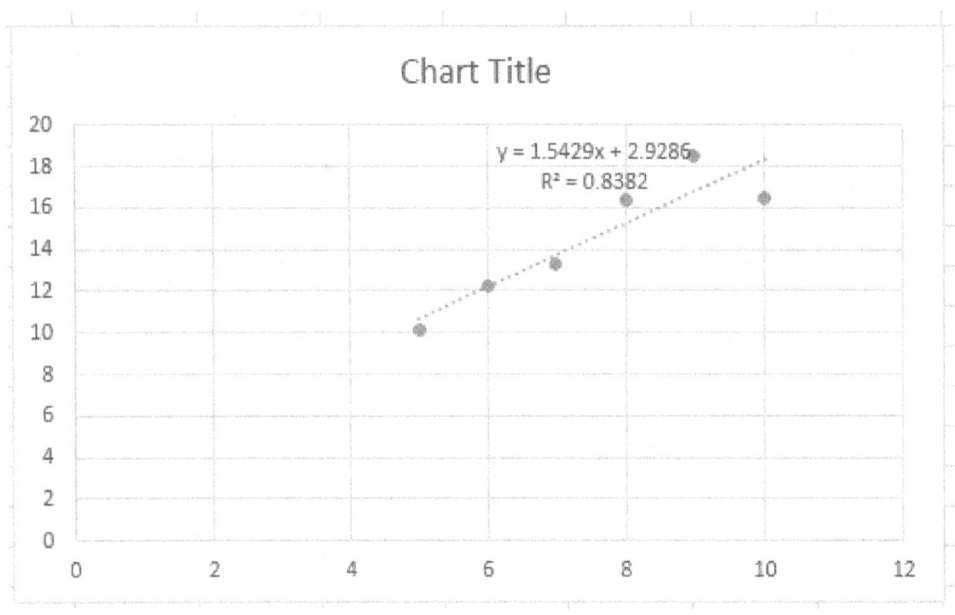

4. r = 0.000000000000

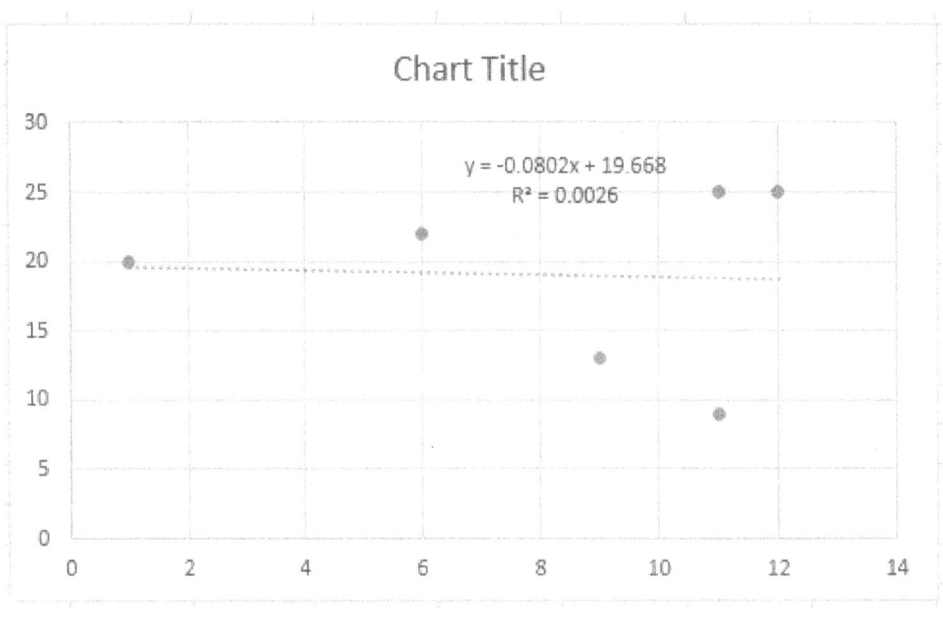

Table of Figures

Figure 1: The Help Function 7
Figure 2: Naming a Cell 8
Figure 3: Cells Highlighted Horizontally 8
Figure 4: Cells Highlighted Vertically 9
Figure 5: Syntax 10
Figure 6: Changing Format of Number 10
Figure 7: Format Cells 11
Figure 8: Format Cell Dialogue Box 11
Figure 9: Adding Analysis Toolpak 12
Figure 10: Adding Analysis Toolpak 2 13
Figure 11: Adding Analysis Toolpak 3 14
Figure 12: Adding Analysis Toolpak 3 14
Figure 13: Data Analysis From Menu Bar 15
Figure 14: Statistical Resources 17
Figure 15: Sorting Data 19
Figure 16: Example Column Chart 20
Figure 17: Example Line Chart 20
Figure 18: Example Pie Chart 21
Figure 19: Example Bar Chart 21
Figure 20: Example Scatter Plot 22
Figure 21: Errors Per Clerk 23
Figure 22: Setup of Frequency Distribution – Part I 24
Figure 23: Setup of Frequency Distribution – Part II 24
Figure 24: Setup of Frequency Distribution – Part III 25
Figure 25: Setup of Frequency Distribution – Part IV 26
Figure 26: Setup of Frequency Distribution – Part V 27
Figure 27: Setup of Frequency Distribution – Part VI 27
Figure 28: Setup of Frequency Distribution – Part VII 28
Figure 29: Setup of Frequency Distribution – Part VIII 29
Figure 30: Test Grades on Statistics Exam 30
Figure 31: Descriptive Stats I 31
Figure 32: Descriptive Stats II 32
Figure 33: Descriptive Stats III 33
Figure 34: Descriptive Stats IV 34
Figure 35: Descriptive Stats V 35
Figure 36: Student Grades 37
Figure 37: Difference From the Mean 38
Figure 38: Differences Squared 38
Figure 39: Sum of Differences Squared 39
Figure 40: Square Root of Variance (Standard Deviation) 40
Figure 41: Standard Error of the Mean 41
Figure 42: Binomial Distribution 54
Figure 43: Components of the Binomial Equation 55
Figure 44: Binomial Setup Part I 56

Figure	Title	Page
Figure 45:	Binomial Setup Part II	57
Figure 46:	Binomial Setup Part II	57
Figure 47:	Binomial Setup Part II	57
Figure 48:	Binomial Setup Part III	58
Figure 49:	Binomial Setup Part IV	59
Figure 50:	Binomial Setup Part V	59
Figure 51:	Binomial Setup Part VI	60
Figure 52:	Binomial Setup Part VI	61
Figure 53:	Binomial Setup Part VII	61
Figure 54:	Sum of Number of Expected Observations	62
Figure 55:	Poisson Setup I	67
Figure 56:	Poisson Setup II	68
Figure 57:	Poisson Setup III	68
Figure 58:	Poisson Example Output	69
Figure 59:	Poisson Data	69
Figure 60:	Poisson Dialogue Box	70
Figure 61:	Poisson Probability of Zero	70
Figure 62:	Poisson Probability of Zero II	71
Figure 63:	Poisson Total Break Summary	71
Figure 64:	Areas of Rejection and Acceptance	77
Figure 65:	Errors Associated With Inferential Statistics	77
Figure 66:	Shape of Normal Distribution	79
Figure 67:	Mean ± 3 SD	80
Figure 68:	Z Score Step 1	81
Figure 69:	Z Score Step 2	81
Figure 70:	Z Score Step 3	82
Figure 71:	Z Score Step 4	82
Figure 72:	Z Score Calculation Example I	84
Figure 73:	Z Score Calculation Example II	84
Figure 74:	Z Score Calculation Example III	84
Figure 75:	Z Score Calculation Example IV	85
Figure 76:	Area Under Curve > and < Than Mean	86
Figure 77:	Small Area To Right Of Data Point	87
Figure 78:	NORMDIST Formula	88
Figure 79:	Determination of Probability of Data Point Occurring Between 2 Z Scores	89
Figure 80:	Confidence Interval	96
Figure 81:	Weights of Meat	96
Figure 82:	Descriptive Stats of Weights of Meat	97
Figure 83:	Descriptive Stats - Weight of Meat Output	98
Figure 84:	Normal vs. t-Distribution	104
Figure 85:	Difference in Appearance Between t-distributions	104
Figure 86:	Confidence Interval for t-distribution	105
Figure 87:	Initial CONFIDENCE.T	106
Figure 88:	Sample of Rejection Area	109
Figure 89:	Probability of Type I Error	110
Figure 90:	Calculation of Z Score	110

Figure	Title	Page
Figure 91:	Customer Wait Time Descriptive Stats	111
Figure 92:	Using Table I	112
Figure 93:	Z Score Formula	113
Figure 94:	Critical vs. Calculated Value	113
Figure 95:	Non-directional Alternative Hypothesis	114
Figure 96:	t-Statistic Calculation	117
Figure 97:	Basketball Points Scored	118
Figure 98:	Areas of Rejection of Null Hypothesis	118
Figure 99:	Critical Value from Table 2	119
Figure 100:	Descriptive Stats	120
Figure 101:	Descriptive Stats II	120
Figure 102:	t-Statistic Calculation	120
Figure 103:	Estimated Location of t-Statistic	121
Figure 104:	Basic Two Sample Flowchart	124
Figure 105:	Z Test Calculation	125
Figure 106:	Z Statistic Calculation	125
Figure 107:	Z-Test - 2 Sample for Means	127
Figure 108:	Z-Test 2 Sample Input	127
Figure 109:	Z-Test 2 Sample Output	128
Figure 110:	Ages of Two Groups	129
Figure 111:	Z-Test 2 Sample Output	130
Figure 112:	Z-Test 2 Sample Output	131
Figure 113:	Comparison of Test Scores	136
Figure 114:	Comparison of Test Scores II	136
Figure 115:	t-Statistic Calculation	137
Figure 116:	t-Statistic Calculation II	137
Figure 117:	Paired t Test	138
Figure 118:	Paired t Test II	138
Figure 119:	Paired t Test Output	139
Figure 120:	Data for F-Test Comparison	145
Figure 121:	F Test Setup	146
Figure 122:	F Test Output	146
Figure 123:	t Statistic and S^2_p Calculation	147
Figure 124:	Sample Data	148
Figure 125:	Pooled Variance Calculation	148
Figure 126:	t Statistic Calculation	148
Figure 127:	Pooled Variance Data	149
Figure 128:	t-Test: Two Sample Assuming Equal Variances Setup	150
Figure 129:	t-Test: Two Sample Assuming Equal Variances Output	150
Figure 130:	Independent Samples with Unequal Variances	151
Figure 131:	Independent Samples with Unequal Variances Setup	151
Figure 132:	Independent Samples with Unequal Variances Output	152
Figure 133:	Variation Sources	161
Figure 134:	Difference Between Distributions I	162
Figure 135:	Difference Between Distributions II	162
Figure 136:	Mean of Squares	164

Figure 137: F Statistic Calculation 164
Figure 138: Example of ANOVA Summary Table 165
Figure 139: Computation of Sum of Squares Step 1 and 2 166
Figure 140: Computation of Sum of Squares and SStot 166
Figure 141: Calculation of $SS_{between}$ 167
Figure 142: Calculation of SS_{within} 167
Figure 143: Tukey's HSD Formula 175
Figure 144: Tukey HSD Dataset 175
Figure 145: ANVOA Output 176
Figure 146: Calculation of Tukey HSD 177
Figure 147: Fisher's Least Significant Difference (LSD) Test 177
Figure 148: Fisher's LSD Data 178
Figure 149: ANOVA Output 178
Figure 150: Fisher's LSD Comparisons 179
Figure 151: Ole Miss and MSU Fans 184
Figure 152: Chi Square ($\chi 2$) Statistic 185
Figure 153: Hours TV Watched vs. GPA 185
Figure 154: Total Rows and Columns 186
Figure 155: Expected Values for Cell I 186
Figure 156: Expected Values for Cell II 186
Figure 157: Arranging Data - Fisher's Exact Test 187
Figure 158: Chisqtest Dialogue Box 187
Figure 159: Highlighted Range in Chitest Dialogue Box 188
Figure 160: GraphPad 2x2 Contingency Table 189
Figure 161: Positive Correlation 194
Figure 162: Negative Correlation 195
Figure 163: Zero Correlation 195
Figure 164: Pearson r Formula 196
Figure 165: Pearson r Interpretation 196
Figure 166: General Equation for a Straight Line 197
Figure 167: Least Sqaures Line 197
Figure 168: r Squared Dataset 198
Figure 169: r Squared Setup I 199
Figure 170: r Squared Setup II 199
Figure 171: r Squared Output 200
Figure 172: r Squared Using CORREL I 200
Figure 173: r Squared Using CORREL II 201
Figure 174: Scatterplot to Obtain r Squared 202
Figure 175: r Squared Using Scatterplot I 202
Figure 176: r Squared Using Scatterplot II 203

REFERENCE: STATISTICAL FUNCTIONS USING MICROSOFT® EXCEL

FUNCTION	DESCRIPTION
AVEDEV function	Returns the average of the absolute deviations of data points from their mean
AVERAGE function	Returns the average of its arguments
AVERAGEA function	Returns the average of its arguments, including numbers, text, and logical values
AVERAGEIF function	Returns the average (arithmetic mean) of all the cells in a range that meet a given criteria
AVERAGEIFS function	Returns the average (arithmetic mean) of all cells that meet multiple criteria
BETA.DIST function	Returns the beta cumulative distribution function
BETA.INV function	Returns the inverse of the cumulative distribution function for a specified beta distribution
BINOM.DIST function	Returns the individual term binomial distribution probability
BINOM.DIST.RANGE function	Returns the probability of a trial result using a binomial distribution
BINOM.INV function	Returns the smallest value for which the cumulative binomial distribution is less than or equal to a criterion value
CHISQ.DIST function	Returns the cumulative beta probability density function
CHISQ.DIST.RT function	Returns the one-tailed probability of the chi-squared distribution
CHISQ.INV function	Returns the cumulative beta probability density function
CHISQ.INV.RT function	Returns the inverse of the one-tailed probability of the chi-squared distribution
CHISQ.TEST function	Returns the test for independence
CONFIDENCE.NORM function	Returns the confidence interval for a population mean
CONFIDENCE.T function	Returns the confidence interval for a population mean, using a Student's t distribution
CORREL function	Returns the correlation coefficient between two data sets
COUNT function	Counts how many numbers are in the list of arguments
COUNTA function	Counts how many values are in the list of arguments
COUNTBLANK function	Counts the number of blank cells within a range
COUNTIF function	Counts the number of cells within a range that meet the given criteria

Function	Description
COUNTIFS function	Counts the number of cells within a range that meet multiple criteria
COVARIANCE.P function	Returns covariance, the average of the products of paired deviations
COVARIANCE.S function	Returns the sample covariance, the average of the products deviations for each data point pair in two data sets
DEVSQ function	Returns the sum of squares of deviations
EXPON.DIST function	Returns the exponential distribution
F.DIST function	Returns the F probability distribution
F.DIST.RT function	Returns the F probability distribution
F.INV function	Returns the inverse of the F probability distribution
F.INV.RT function	Returns the inverse of the F probability distribution
F.TEST function	Returns the result of an F-test
FISHER function	Returns the Fisher transformation
FISHERINV function	Returns the inverse of the Fisher transformation
FORECAST function	Returns a value along a linear trend
FREQUENCY function	Returns a frequency distribution as a vertical array
GAMMA function	Returns the gamma function value
GAMMA.DIST function	Returns the gamma distribution
GAMMA.INV function	Returns the inverse of the gamma cumulative distribution
GAMMALN function	Returns the natural logarithm of the gamma function, $\Gamma(x)$
GAMMALN.PRECISE function	Returns the natural logarithm of the gamma function, $\Gamma(x)$
GAUSS function	Returns 0.5 less than the standard normal cumulative distribution
GEOMEAN function	Returns the geometric mean
GROWTH function	Returns values along an exponential trend
HARMEAN function	Returns the harmonic mean
HYPGEOM.DIST function	Returns the hypergeometric distribution
INTERCEPT function	Returns the intercept of the linear regression line
KURT function	Returns the kurtosis of a data set
LARGE function	Returns the k-th largest value in a data set
LINEST function	Returns the parameters of a linear trend
LOGEST function	Returns the parameters of an exponential trend
LOGNORM.DIST function	Returns the cumulative lognormal distribution
LOGNORM.INV function	Returns the inverse of the lognormal cumulative distribution

Function	Description
MAX function	Returns the maximum value in a list of arguments
MAXA function	Returns the maximum value in a list of arguments, including numbers, text, and logical values
MEDIAN function	Returns the median of the given numbers
MIN function	Returns the minimum value in a list of arguments
MINA function	Returns the smallest value in a list of arguments, including numbers, text, and logical values
MODE.MULT function	Returns a vertical array of the most frequently occurring, or repetitive values in an array or range of data
MODE.SNGL function	Returns the most common value in a data set
NEGBINOM.DIST function	Returns the negative binomial distribution
NORM.DIST function	Returns the normal cumulative distribution
NORM.INV function	Returns the inverse of the normal cumulative distribution
NORM.S.DIST function	Returns the standard normal cumulative distribution
NORM.S.INV function	Returns the inverse of the standard normal cumulative distribution
PEARSON function	Returns the Pearson product moment correlation coefficient
PERCENTILE.EXC function	Returns the k-th percentile of values in a range, where k is in the range 0..1, exclusive.
PERCENTILE.INC function	Returns the k-th percentile of values in a range
PERCENTRANK.EXC function	Returns the rank of a value in a data set as a percentage (0..1, exclusive) of the data set
PERCENTRANK.INC function	Returns the percentage rank of a value in a data set
PERMUT function	Returns the number of permutations for a given number of objects
PERMUTATIONA function	Returns the number of permutations for a given number of objects (with repetitions) that can be selected from the total objects
PHI function	Returns the value of the density function for a standard normal distribution
POISSON.DIST function	Returns the Poisson distribution
PROB function	Returns the probability that values in a range are between two limits
QUARTILE.EXC function	Returns the quartile of the data set, based on percentile values from 0..1, exclusive
QUARTILE.INC function	Returns the quartile of a data set
RANK.AVG function	Returns the rank of a number in a list of numbers

RANK.EQ function	Returns the rank of a number in a list of numbers
RSQ function	Returns the square of the Pearson product moment correlation coefficient
SKEW function	Returns the skewness of a distribution
SKEW.P function	Returns the skewness of a distribution based on a population: a characterization of the degree of asymmetry of a distribution around its mean
SLOPE function	Returns the slope of the linear regression line
SMALL function	Returns the k-th smallest value in a data set
STANDARDIZE function	Returns a normalized value
STDEV.P function	Calculates standard deviation based on the entire population
STDEV.S function	Estimates standard deviation based on a sample
STDEVA function	Estimates standard deviation based on a sample, including numbers, text, and logical values
STDEVPA function	Calculates standard deviation based on the entire population, including numbers, text, and logical values
STEYX function	Returns the standard error of the predicted y-value for each x in the regression
T.DIST function	Returns the Percentage Points (probability) for the Student t-distribution
T.DIST.2T function	Returns the Percentage Points (probability) for the Student t-distribution
T.DIST.RT function	Returns the Student's t-distribution
T.INV function	Returns the t-value of the Student's t-distribution as a function of the probability and the degrees of freedom
T.INV.2T function	Returns the inverse of the Student's t-distribution
T.TEST function	Returns the probability associated with a Student's t-test
TREND function	Returns values along a linear trend
TRIMMEAN function	Returns the mean of the interior of a data set
VAR.P function	Calculates variance based on the entire population
VAR.S function	Estimates variance based on a sample
VARA function	Estimates variance based on a sample, including numbers, text, and logical values
VARPA function	Calculates variance based on the entire population, including numbers, text, and logical values
WEIBULL.DIST function	Returns the Weibull distribution
Z.TEST function	Returns the one-tailed probability-value of a z-test

Source: Microsoft® Excel help

TABLE I: The Standard Normal Distribution

z	0	0.01	0.02	0.03	0.04	0.05	0.06	0.07	0.08	0.09
0	0	0.00399	0.00798	0.01197	0.01595	0.01994	0.02392	0.0279	0.03188	0.03586
0.1	0.0398	0.0438	0.04776	0.05172	0.05567	0.05966	0.0636	0.06749	0.07142	0.07535
0.2	0.0793	0.08317	0.08706	0.09095	0.09483	0.09871	0.10257	0.10642	0.11026	0.11409
0.3	0.11791	0.12172	0.12552	0.1293	0.13307	0.13683	0.14058	0.14431	0.14803	0.15173
0.4	0.15542	0.1591	0.16276	0.1664	0.17003	0.17364	0.17724	0.18082	0.18439	0.18793
0.5	0.19146	0.19497	0.19847	0.20194	0.2054	0.20884	0.21226	0.21566	0.21904	0.2224
0.6	0.22575	0.22907	0.23237	0.23565	0.23891	0.24215	0.24537	0.24857	0.25175	0.2549
0.7	0.25804	0.26115	0.26424	0.2673	0.27035	0.27337	0.27637	0.27935	0.2823	0.28524
0.8	0.28814	0.29103	0.29389	0.29673	0.29955	0.30234	0.30511	0.30785	0.31057	0.31327
0.9	0.31594	0.31859	0.32121	0.32381	0.32639	0.32894	0.33147	0.33398	0.33646	0.33891
1	0.34134	0.34375	0.34614	0.34849	0.35083	0.35314	0.35543	0.35769	0.35993	0.36214
1.1	0.36433	0.3665	0.36864	0.37076	0.37286	0.37493	0.37698	0.379	0.381	0.38298
1.2	0.38493	0.38686	0.38877	0.39065	0.39251	0.39435	0.39617	0.39796	0.39973	0.40147
1.3	0.4032	0.4049	0.40658	0.40824	0.40988	0.41149	0.41308	0.41466	0.41621	0.41774
1.4	0.41924	0.42073	0.4222	0.42364	0.42507	0.42647	0.42785	0.42922	0.43056	0.43189
1.5	0.43319	0.43448	0.43574	0.43699	0.43822	0.43943	0.44062	0.44179	0.44295	0.44408
1.6	0.4452	0.4463	0.44738	0.44845	0.4495	0.45053	0.45154	0.45254	0.45352	0.45449
1.7	0.45543	0.45637	0.45728	0.45818	0.45907	0.45994	0.4608	0.46164	0.46246	0.46327
1.8	0.46407	0.46485	0.46562	0.46638	0.46712	0.46784	0.46856	0.46926	0.46995	0.47062
1.9	0.47128	0.47193	0.47257	0.4732	0.47381	0.47441	0.475	0.47558	0.47615	0.4767
2	0.47725	0.47778	0.47831	0.47882	0.47932	0.47982	0.4803	0.48077	0.48124	0.48169
2.1	0.48214	0.48257	0.483	0.48341	0.48382	0.48422	0.48461	0.485	0.48537	0.48574
2.2	0.4861	0.48645	0.48679	0.48713	0.48745	0.48778	0.48809	0.4884	0.4887	0.48899
2.3	0.48928	0.48956	0.48983	0.4901	0.49036	0.49061	0.49086	0.49111	0.49134	0.49158
2.4	0.4918	0.49202	0.49224	0.49245	0.49266	0.49286	0.49305	0.49324	0.49343	0.49361
2.5	0.49379	0.49396	0.49413	0.4943	0.49446	0.49461	0.49477	0.49492	0.49506	0.4952
2.6	0.49534	0.49547	0.4956	0.49573	0.49585	0.49598	0.49609	0.49621	0.49632	0.49643
2.7	0.49653	0.49664	0.49674	0.49683	0.49693	0.49702	0.49711	0.4972	0.49728	0.49736
2.8	0.49744	0.49752	0.4976	0.49767	0.49774	0.49781	0.49788	0.49795	0.49801	0.49807
2.9	0.49813	0.49819	0.49825	0.49831	0.49836	0.49841	0.49846	0.49851	0.49856	0.49861
3	0.49865	0.49869	0.49874	0.49878	0.49882	0.49886	0.49889	0.49893	0.49896	0.499

Source: Wikipedia

TABLE 2: Student's t Distribution

1 Sided	75%	80%	85%	90%	95%	97.50%	99%	99.50%	99.75%	99.90%	99.95%
2 Sided	50%	60%	70%	80%	90%	95%	98%	99%	99.50%	99.80%	99.90%
1	1	1.376	1.963	3.078	6.314	12.71	31.82	63.66	127.3	318.3	636.6
2	0.816	1.08	1.386	1.886	2.92	4.303	6.965	9.925	14.09	22.33	31.6
3	0.765	0.978	1.25	1.638	2.353	3.182	4.541	5.841	7.453	10.21	12.92
4	0.741	0.941	1.19	1.533	2.132	2.776	3.747	4.604	5.598	7.173	8.61
5	0.727	0.92	1.156	1.476	2.015	2.571	3.365	4.032	4.773	5.893	6.869
6	0.718	0.906	1.134	1.44	1.943	2.447	3.143	3.707	4.317	5.208	5.959
7	0.711	0.896	1.119	1.415	1.895	2.365	2.998	3.499	4.029	4.785	5.408
8	0.706	0.889	1.108	1.397	1.86	2.306	2.896	3.355	3.833	4.501	5.041
9	0.703	0.883	1.1	1.383	1.833	2.262	2.821	3.25	3.69	4.297	4.781
10	0.7	0.879	1.093	1.372	1.812	2.228	2.764	3.169	3.581	4.144	4.587
11	0.697	0.876	1.088	1.363	1.796	2.201	2.718	3.106	3.497	4.025	4.437
12	0.695	0.873	1.083	1.356	1.782	2.179	2.681	3.055	3.428	3.93	4.318
13	0.694	0.87	1.079	1.35	1.771	2.16	2.65	3.012	3.372	3.852	4.221
14	0.692	0.868	1.076	1.345	1.761	2.145	2.624	2.977	3.326	3.787	4.14
15	0.691	0.866	1.074	1.341	1.753	2.131	2.602	2.947	3.286	3.733	4.073
16	0.69	0.865	1.071	1.337	1.746	2.12	2.583	2.921	3.252	3.686	4.015
17	0.689	0.863	1.069	1.333	1.74	2.11	2.567	2.898	3.222	3.646	3.965
18	0.688	0.862	1.067	1.33	1.734	2.101	2.552	2.878	3.197	3.61	3.922
19	0.688	0.861	1.066	1.328	1.729	2.093	2.539	2.861	3.174	3.579	3.883
20	0.687	0.86	1.064	1.325	1.725	2.086	2.528	2.845	3.153	3.552	3.85
21	0.686	0.859	1.063	1.323	1.721	2.08	2.518	2.831	3.135	3.527	3.819
22	0.686	0.858	1.061	1.321	1.717	2.074	2.508	2.819	3.119	3.505	3.792
23	0.685	0.858	1.06	1.319	1.714	2.069	2.5	2.807	3.104	3.485	3.767
24	0.685	0.857	1.059	1.318	1.711	2.064	2.492	2.797	3.091	3.467	3.745
25	0.684	0.856	1.058	1.316	1.708	2.06	2.485	2.787	3.078	3.45	3.725
26	0.684	0.856	1.058	1.315	1.706	2.056	2.479	2.779	3.067	3.435	3.707
27	0.684	0.855	1.057	1.314	1.703	2.052	2.473	2.771	3.057	3.421	3.69
28	0.683	0.855	1.056	1.313	1.701	2.048	2.467	2.763	3.047	3.408	3.674
29	0.683	0.854	1.055	1.311	1.699	2.045	2.462	2.756	3.038	3.396	3.659
30	0.683	0.854	1.055	1.31	1.697	2.042	2.457	2.75	3.03	3.385	3.646
40	0.681	0.851	1.05	1.303	1.684	2.021	2.423	2.704	2.971	3.307	3.551
50	0.679	0.849	1.047	1.299	1.676	2.009	2.403	2.678	2.937	3.261	3.496
60	0.679	0.848	1.045	1.296	1.671	2	2.39	2.66	2.915	3.232	3.46
80	0.678	0.846	1.043	1.292	1.664	1.99	2.374	2.639	2.887	3.195	3.416
100	0.677	0.845	1.042	1.29	1.66	1.984	2.364	2.626	2.871	3.174	3.39
120	0.677	0.845	1.041	1.289	1.658	1.98	2.358	2.617	2.86	3.16	3.373
∞	0.674	0.842	1.036	1.282	1.645	1.96	2.326	2.576	2.807	3.09	3.291

Source: Wikipedia

INDEX

Analysis of Variance, 161
Analysis Toolpak, 12
ANOVA summary table, 166
ANOVA: two factor with replication, 169
ANOVA: two factor without replication, 169
area under the normal curve, 89
binomial distribution, 53
binominal equation, 54
BINS, 23
central limit theorem, 96
change the number of decimal places, 10
charts and graphs, 19
chi square (χ^2) statistic, 186
chi square test, 186
classes of correlation, 195
coefficient of determination, 199
coefficient of variation, 41
column, 7
comparison of two samples, 125
confidence intervals, 96
contingency table, 185
Correlation, 195
Count, 36
critical value, 77
data, 9
Data Analysis, 15
degrees of freedom, 104
descriptive statistics, 30
difference between the t distribution and Normal distribution, 104
differences between measured values, 110
discrete data, 53
experimental error, 163
F distribution, 165
File, 9
Fisher's Exact Test, 189
Fisher's Least Significant Difference (LSD) test, 178
format cells, 11
formatting of a cell, 10
formulas, 9
frequency distribution, 22
graphical representations, 18
home, 9
hypothesis, 77
inferential statistics, 76
inherent variation, 163
Kurtosis, 36
linear regression, 199
mathematical functions, 6, 9
Maximum, 36
Mean, 35
menu bar, 9
Microsoft® Excel, 6
Microsoft® Excel's help function, 6
Minimum, 36
Mode, 36
nonrejection, 77
normal deviate, 97
normal distribution, 79
normally distributed curve, 80
null hypothesis, 77
paired t test, 136
parameter, 77
Pearson r, 197
point estimates, 96
poisson distribution, 67
poisson distribution formula, 67
pooled variance t-test, 150
post hoc tests, 175
probability, 54
proportions - differences, 185
r^2, 199
Range, 36
relative frequency, 28
row, 7
significance in small samples, 118
Skewness, 36
Standard Deviation, 36
standard deviation - binomial, 62
Standard Error, 35
Standard Error of the Mean, 40
standard normal deviate (Z), 81
statistic, 76
statistical functions - excel reference, 213
statistical packages, 16
straight line - equation, 198
Sum, 36

sum of squares, 39
syntax, 10
t distribution, 104
t Distribution - table, 220
Table - Standard Normal Distribution, 218
test statistic, 77
treatment effect, 164
t-Test: Two-Sample Assuming Equal Variances, 150
t-tests of equal and unequal variances, 145
Turkey's HSD, 175
two sample flowchart, 125
type I error, 78
type II error, 78
Variance, 36
Z score, 82

www.ingramcontent.com/pod-product-compliance
Lightning Source LLC
Chambersburg PA
CBHW080909170526
45158CB00008B/2046